PRINCIPLES OF AQUACULTURE
Theory Manual

THE AUTHORS

M. Porkodi working as an Assistant Professor in the Department of Aquaculture in the College of Fisheries Science, St.Devasahayam Institute of Fisheries Science and Technology, Midalam, Kanyakumari district. She has completed her UG degree at Dr.M.G.R Fisheries College and Research Institute, Ponneri and Master's degree at ICAR-Central Institute of Fisheries Education, Mumbai. She was awarded with ICAR-JRF Fellowship during her Master's degree Programme. She has successfully carried out a Project with the title of " Estimation of Biogrowth parameters of Mystus aor fish species through floating fish feed". Additionally, she has successfully conducted a few capacity-building training program with funding obtained from NFDB, Hyderabad.

Prof. S. Felix is a distinguished figure in the field of fisheries education and technology consulting, with a career spanning over four decades. Currently serving as the Dean of the College of Fisheries Science at DIFST (Devasahayam Institute of Fisheries Science and Technology) in Midalam, Kanniyakumari district, Tamil Nadu and also President of AQUAFIN (Aquaculture Technology & Research Foundation India), he has held several prestigious positions, including Vice Chancellor of Tamil Nadu Fisheries University (TNJFU) and President of the World Aquaculture Society (WAS)-APC. Prof. Felix is recognized for his exemplary leadership and contributions to the fisheries and aquaculture sector, demonstrated through his involvement in various committees and organizations. He is a prolific author, having published numerous research papers, books, and manuals, and has played a pivotal role in introducing new fisheries science courses and establishing fisheries colleges and research centers.

PRINCIPLES OF AQUACULTURE
Theory Manual

M. Porkodi
S. Felix

College of Fisheries Science
St. Devasahayam Institute of
Fisheries Science & Technology
Midalam, Kanniyakumari Dist.629193 (Tamil Nadu)

2025

Daya Publishing House®
A Division of
Astral International Pvt. Ltd.

ISBN: 978-93-5919-515-5

Published by : **Daya Publishing House®**
A Division of
Astral International Pvt. Ltd.
– ISO 9001:2015 Certified Company –
4736/23, Ansari Road, Darya Ganj
New Delhi-110 002
Ph. 011-43549197, 23278134
E-mail: info@astralint.com
Website: www.astralint.com

Printed at : **Replika Press Pvt. Ltd**

Preface

Aquaculture, the practice of culturing aquatic organisms, has emerged as a pivotal sector in addressing global food security, economic growth, and environmental sustainability. With the world's population continuously expanding and natural fisheries facing unprecedented pressures, aquaculture offers a promising solution to meet the increasing demand for seafood while mitigating the impacts on wild fish stocks. "Principles of Aquaculture" is a comprehensive guide designed to provide both novice enthusiasts and seasoned practitioners with a solid foundation in the fundamental principles, techniques, and management strategies essential for successful aquaculture operations. This book is crafted to serve as an invaluable resource for students, researchers, industry professionals, policymakers, and anyone seeking to understand and engage in the dynamic field of aquaculture. Drawing upon the latest research findings, technological advancements, and practical experiences from experts in the field, Principles of Aquaculture covers a wide range of topics including the biology and ecology of aquatic organisms, aquaculture systems and infrastructure, water quality management, nutrition and feeding regimes, disease prevention and control, genetics and selective breeding, as well as economic and environmental considerations. Each chapter is meticulously crafted to provide clear explanations, supported by illustrative diagrams, tables, and case studies, facilitating a deeper understanding of the subject matter. Moreover, emphasis is placed on the interdisciplinary nature of aquaculture, highlighting its connections with marine biology, environmental science, engineering, economics, and social sciences, among others. In an era marked by increasing concerns over food security, environmental degradation, and climate change, the sustainable development of aquaculture is paramount. Thus, this book not only addresses the technical aspects of aquaculture but also underscores the importance of adopting environmentally responsible practices, promoting social equity, and ensuring economic viability.

We thank our Students, Shiyoba, Arthi, Raksha, Asvitha and Delfia for helping us in making the drawing for this manual.

M. Porkodi

S. Felix

Contents

Chapter-1

Aquaculture Basics and Scope

Aquaculture stands as a significant and expanding sector, playing a crucial role in producing high-protein food. In 2020, global aquaculture yielded 178 million tons of food fish, including finfish, crustaceans, mollusks, and other aquatic animals, for human consumption (SOFIA, 2020). The share of aquaculture in total fisheries and aquaculture production has risen steadily from 34.5 percent in 2006 to 62.2 percent in 2020. Asia has emerged as a dominant force in global aquaculture, accounting for 88.8% of production quantity and 78.7% of production value in 2008. China led the world in aquaculture production with 9.4%, followed by India with 7.1% (2008).

The Food and Agriculture Organization of the United Nations (FAO) defines aquaculture as the farming of aquatic organisms, including fish, mollusks, crustaceans, and aquatic plants. Farming involves interventions such as regular feeding, stocking, and predator protection to enhance productivity. Aquaculture specifically refers to the harvesting of aquatic organisms by individuals or corporate entities who own them throughout their rearing period, for statistical purposes (FAO, 1997b).

Advancements in aquaculture technology allow for profitable commercial production of various species by managing their entire life cycles. Controlled production of "seed" materials, including larvae and juveniles, eliminates the need to collect them from the wild, starting from the development of broodstock. Understanding each species' behavior, habitat requirements, reproductive biology, dietary needs, and susceptibility to diseases in culture conditions is essential for successful aquaculture. This involves creating facilities for various life-cycle stages, such as grow-out facilities, feed development, fish handling systems, disease management, broodstock holding tanks, and sea cages.

The field of fisheries in Indian agriculture is rapidly growing, offering a plethora of methodologies and techniques for marine and freshwater fish species as well as shellfish. This sector holds enormous potential to diversify farming practices, uplift rural livelihoods, bolster domestic nutritional security, generate employment opportunities, increase export revenues, and even stimulate tourism. From expansive seas to high-altitude mountainous regions, including valuable cold-water species, there are untapped opportunities across island systems, ranging from ornamental fish to value-added products. Presently, approximately 5 million fisherfolk are employed from coastal villages, major river basins, and reservoirs nationwide. Over the last five decades, India has witnessed remarkable growth in fish production, soaring from 0.75 million tons in the 1950s to 6.3 million tons in 2004, with particularly noteworthy progress since 2001-2002. In 2002-03, the fisheries sector contributed Rs 268,500 million to the Gross Domestic Product (GDP), constituting 1.19% of the total GDP. With an extensive coastline stretching 8,129 km and a vast Exclusive Economic Zone (EEZ) spanning 2.02 million sq km, including areas around the Andaman and Nicobar Islands and the continental shelf, India holds an annual marine potential of 3.93 million tons. Furthermore, the nation boasts extensive brackish water areas along the coast suitable for sea farming and mariculture.

Objectives of Aquaculture

- Generating affordable, protein-rich, nutritious, flavorful, and easily digestible food for human consumption.
- Enhancing fish populations in both natural and artificial water sources through artificial breeding and stocking.
- Cultivating ornamental fish species to enhance visual aesthetics.
- Maximizing the efficient use of aquatic and terrestrial resources.
- Recycling organic waste from human and livestock sources.
- Creating employment opportunities through commercial and industrial aquaculture.
- Cultivating sportfish and supporting recreational fishing activities.
- Producing bait-fish for both commercial and recreational fishing purposes.

In aquaculture, waters are categorized into three types: freshwater, saltwater, and brackish. Freshwaters, typically found inland, and the saltwater of oceans and seas, exhibit significant differences in salinity, ranging from negligible in freshwater to nearly 35 parts per thousand (ppt) in the oceans. The salinity of these waters plays a crucial role in shaping the aquatic life they support. For instance, certain fish and shellfish species are stenohaline, meaning they can only tolerate narrow salinity ranges. Carp exemplifies stenohaline freshwater fish, while sardines or mackerels represent stenohaline saltwater fish.

Brackish water naturally occurs in estuaries, river deltas, lagoons, and backwaters, subject to tidal fluctuations. Salinity in these habitats varies widely, from 0 to 35 ppt, influenced by tides and freshwater inflow. Species inhabiting brackish waters, such as mullet and mud-skippers among fish, and various penaeid prawns and crabs among crustaceans, are typically euryhaline, capable of withstanding broad salinity changes. Some fish and shellfish species, termed anadromous or catadromous, move between freshwater streams, estuaries, and the sea during different life stages. Anadromous species like salmon and shad breed in freshwater but spend adulthood in the sea, while catadromous species like eels migrate from rivers to the sea for breeding. Additionally, certain species, such as some palaeomonid prawns, exhibit restricted migration between freshwater river sections and estuaries.

Some species breed in estuaries but spend their mid-years in freshwater environments, while others migrate between estuaries or lagoons and the sea at different life stages. Examples include mullet (such as Mugil cephalus) and shrimp (such as Penaeus monodon, P. notialis), which exhibit such migration patterns.

The significant growth in aquaculture has been driven by declines in wild fish and shellfish stocks. Each aquaculture system requires specific conditions, resulting in varying biomass production. Selection of an aquaculture system depends on factors like geographical location, water type, target species, resource availability, timing of seed availability, and socioeconomic considerations. Fishes dominate aquaculture in terms of species diversity, quantity, and value. Fish culture can be practiced in various water bodies, including freshwater, brackish water, seawater, or even sewage water. Cultivation may focus on a single fish species or utilize compatible species combinations to maximize biomass. Fish can be raised in traditional community ponds without artificial feed supplementation or intensive management practices.

Types of Aquaculture

Various aquaculture methods have evolved, with modern practices emphasizing intensification, characterized by high stocking densities and artificial feed, while closely monitoring water quality and environmental conditions to optimize fish growth. Additionally, aquaculture can be integrated with other agricultural or animal farming systems, such as cultivating fish in rice fields with ample water or utilizing pond bunds for agricultural or animal husbandry purposes.

Scope of Aquaculture Worldwide

According to SOFIA 2022, in 2020, total capture fisheries production reached around 90.3 Million tonnes comprised of 11.5 MT inland and 78.8 MT Marine Capture fisheries Production. Total Aquaculture production in the year 2020 reached around 87.5 MT comprising of 54.4 MT inland Aquaculture fish Production and 33.1 MT of Marine Aquaculture Production. Total **World Fisheries and Aquaculture Production** reached around 177.8 MT.In World **Percapita apparent**

fish consumption reached around 20.2 Kg/Annum in the year 2020. The fisheries and Aquaculture sector not only produce food but also provides livelihood to the 59 million people over 90% of these in small-scale fisheries. According to SOFIA 2022 data, **China** is in the 1st Position in Aquatic animal Production all over the world. India is in **2nd** Position in World Aquatic Animal Production. In inland aquaculture production of finfish, China is in 1st Position and India is in 2nd Position. In marine and coastal aquaculture production of finfish, China is in 1st Position and India is in 13th position. Global apparent consumption of aquatic foods increased at an **average annual rate** of **3.0** percent from 1961 to 2019.

Per capita consumption of aquatic animal foods grew by about 1.4 percent per year, from 9.0 kg (live weight equivalent) in 1961 to 20.5 kg in 2019. Preliminary data for 2020 point to a slight decline to 20.2 kg. In the same year, aquaculture accounted for 56 percent of the amount of aquatic animal food production available for human consumption. During recent decades, per capita consumption of aquatic foods has been influenced most strongly by increased supplies, changing consumer preferences, advancements in technology and income growth.

Scope of Aquaculture in India

Freshwater aquaculture in India predominantly focuses on three major carp species: rohu, catla, and mrigala, alongside exotic species like silver carp, grass carp, and common carp. Among catfish, magur (*Clarias batrachus*) stands out due to its consumer preference, market value, and suitability for farming in adverse ecological conditions. Recent interest has surged in *Pangasius* spp., particularly in the Koleru lake region of Andhra Pradesh, due to its growth potential and market demand. Other potential species include various carps, barbs, and cirrhinus species, albeit currently cultured at low levels based on wild seed collection.

In brackish water aquaculture, Penaeus monodon shrimp culture dominates, while other species like *Fenneropenaeus indicus* and *Marsupenaeus japonicus* are not extensively farmed commercially. Development of Fenneropenaeus Vannamei culture is emerging. Promising marine finfish species like seabass, grouper, grey mullet, and pearl-spot remain underexploited. Mariculture is expected to grow significantly along the Indian coastline, driven by the wide range of cultivable species, available technologies, and favorable climate.

Aquaculture holds immense potential to address food insecurity and supplement animal feed. The sector has evolved into a modern, dynamic industry, focusing on producing safe, high-value products while emphasizing environmental sustainability. Eco-friendly aquaculture practices aligned with societal and environmental needs are imperative for sustainable development in India and globally.

Chapter-2

History of Aquaculture

Theories of Aquaculture

Several theories propose explanations for the origins of aquaculture, including the oxbow theory, catch-and-hold theory, concentration theory, and trap-and-crop theory.

A, Oxbow theory

The origin of aquaculture, as per this narrative, can be traced back to natural rivers and streams, particularly in inland regions. Over time, these water bodies develop meanders and oxbow lakes, which, due to the local topography, may vary in size and shape. Fluctuating flood levels lead to changes in river courses, leaving behind these oxbow lakes along with their fish and other organisms.

Local communities, often reliant on fishing, discovered the abundant fish harvests in these natural oxbow lakes. They observed that seasonal flooding restocked these water bodies with fish, which could be harvested during the dry season. Recognizing the potential, enterprising individuals began to improve embankments, enclosing and managing these oxbow lakes for aquaculture. This process continued until full aquaculture management was achieved.

Such developments typically occur in regions characterized by extensive low-lying areas with a network of rivers and distinct monsoonal seasons, with Bangladesh serving as a prime example of such terrain.

B, Concentration Theory

Tropical regions experience monsoons, characterized by heavy rainfall and floods followed by a dry season. During the rainy period, rivers swell and flood extensive lowland areas. Marshlands, abundant in vegetation and aquatic life, become thriving habitats for fish during this flood season. As the rains subside, water gradually recedes, leaving behind deep spots and rivers.

This leads to the concentration of fish, which have thrived and reproduced during the wet season, in these watered depressions or rivers. Local fishermen capitalize on this concentration to catch fish, initially without discriminating based on size or species. Over time, efforts are made to preserve smaller fish or transfer them to other rearing areas.

In some cases, embankments are constructed around suitable depressions to facilitate fish culture, marking the beginning of aquaculture management through concentration-based practices.

C, Catch and Hold Theory

Early rulers of large empires highly valued fish and aquatic products, often incorporating water bodies around castles for water supply, recreation, and defense. While these water bodies were not initially intended for fish rearing, some rulers demanded fish year-round, prompting officials to devise means to procure fish even in winter. This led to the practice of stocking fish caught from natural waters into these constructed water bodies.

Over time, certain species, such as common carp, thrived in these artificial waters, leading to the establishment of a catch- and-hold system for fish provision. Further advancements included stocking the appropriate quantity and species of fish and providing supplementary feeding when necessary, eventually evolving into actual aquaculture practices. Monasteries in Europe and the palaces of emperors and rulers served as prominent venues for the development of aquaculture during this period.

D, Trap and Crop Theory

The fourth theory of aquaculture development contrasts with inland freshwater methods, focusing on brackish and marine regions influenced by tidal changes. Coastal areas feature various water bodies like coves, lagoons, ponds, and swamps that experience periodic flooding and drainage due to tides. Coastal communities have long recognized these areas as natural habitats for fish, crustaceans, mollusks, and other aquatic resources.

Capitalizing on this natural abundance, fishermen began installing traps to capture fish and crustaceans entering these areas during flood tides. This management strategy ensured a regular supply of fish for personal consumption and sale. Over time, increased fishing activity led to a decline in trapped fish, prompting fishermen to leave the barricades in place for longer periods, typically one to three months, allowing captured fish to grow to marketable sizes.

This chronology characterizes the development of brackish water aquaculture, believed to have originated in Indonesia before spreading to the Philippines, Thailand, Malaysia, India, and beyond.

History of Aquaculture Worldwide

2000-1000 B.C

Aquaculture traces its origins back to approximately 2000-1000 B.C., with Chinese aquaculturists recognized as the earliest pioneers, marking a history of about 4000 years. China is renowned as the birthplace of aquaculture, with a primary emphasis on cultivating the common carp (*Cyprinus carpio*).

500 B.C. (473 B.C. or 475 B.C.)

The year 475 B.C. holds profound historical significance in the development of aquaculture, as it marks the authorship of "The Classic of Fish Culture" by Fan Lai, widely regarded as the earliest monograph on fish culture. While some sources suggest variations such as 500 B.C. or 473 B.C., most agree on the year 475 B.C. This seminal work not only delves into metaphysical aspects but also provides detailed descriptions of pond structures, common carp propagation methods, and fry growth, laying the foundation for early aquacultural practices.

500 B.C.- 500 A.D.

This period can be considered the **Golden Age of common carp culture**, which has continued to develop in China and neighboring countries where Chinese migration or foreign relations have occurred. Significant advancements have been made in the techniques of culture but also scattered records of the culture systems were made during this period. Around the same time, specifically between 321 and 300 B.C., the use of reservoirs to hold fish was first described in the Indian sub-continent.

618-906 A.D (Tang dynasty)

During the Tang Dynasty in China, the reign of Emperor Li brought about a pivotal moment in the history of world aquaculture. Despite an imperial decree banning the culture of the common carp, which shared the emperor's family name, this restriction unexpectedly spurred innovation in aquaculture. Seeking alternatives, the Chinese people discovered the silver carp, big-head carp, grass carp, and mud carp, all well suited for pond culture. When raised together in polyculture within the same pond, these species complemented each other's diets and environmental preferences, significantly enhancing freshwater pond productivity.

906-1900 A.D

906-1120 (Sung Dynasty)

During the Tang Dynasty, a new initiative began to collect fry of cultivable species seasonally along rivers, spurred by the prohibition decree targeting the common carp. This practice saw significant refinement during the Sung Dynasty, with systematic fry collection and dispersal in natural waters becoming highly advanced. Meanwhile, in India, the published work Namasollasa documented the process of fattening fish in reservoirs.

1368-1644 (Ming Dynasty)

In the Ming period, comprehensive manuals outlining the entire aquaculture process emerged, covering aspects such as fry cultivation to adulthood, pond structures, stocking densities, polyculture techniques, rotation of stocking and harvesting, application of feed and fertilizer, and disease management. Additionally, brackish water aquaculture was documented to have begun in Indonesia around 1400.

1644 to 1911 (Ching Dynasty)

During this era, there was a heightened focus on providing detailed descriptions of fish culture methods, encompassing fry production, the timingof fry occurrence, fry differentiation, separation techniques, and transportation methods.

Development in Other Areas

French Indochina

The influence of Chinese migration waves in the French Indochinese countries contributed to the development of aquaculture in the region. Cage culture of siluroids and related species emerged independently, driven by the presence of indigenous species. This distinctive aquaculture practice has persisted to the present day, as evidenced by ongoing cage culture along the Mekong River in Kampuchea.

Sub Continent of India

The tradition of constructing water reservoirs of different sizes for water supply and religious functions began in this region during ancient times. Initially, these reservoirs were not utilized for fish culture. However, over time, they transitioned from merely holding fish to actively cultivating them.

Indonesia

This country is credited with pioneering the early development of brackish water aquaculture at the onset of the 15th century. This initiative subsequently spread to neighboring regions, including the Philippines, Malaysia, Thailand, and the southern parts of China (Taiwan).

Europe

Aquaculture in Europe has its origins dating back to ancient times. Water areas were initially established in the palaces of early rulers, as well as in temples and monasteries. Initially serving decorative and religious purposes, these water bodies later transitioned into temporary holding areas for fish. Eventually, they evolved into environments suitable for fish culture, with common carp and trout emerging as the primary species cultivated.

North America

During the 19th century, efforts to advance aquaculture, particularly focused on promoting sport fishing, were undertaken. The United States Commission of Fish and Fisheries published "A Manual of Fish Culture" in 1897, which primarily addressed the establishment of hatcheries for seed production to stock game waters. Additionally, it covered the cultivation of some food species of finfish, oysters, clams, and other aquatic organisms.

Africa

Earlier efforts, primarily from Europe, aimed to introduce aquaculture practices in African countries, leveraging the presence of extensive floodplains conducive to the growth and reproduction of indigenous species. This laid the groundwork for the early development of aquaculture on the continent. Presently, numerous initiatives for aquaculture development are underway across various African nations, focusing on species such as tilapia, common carp, and catfish.

Initially, seeds or fry were collected from wild waters for aquaculture. The expansion of the aquaculture industry necessitated breakthroughs in technology for induced spawning of cultivable species between 1700 and 1900. During this period, Asian carps, Indian major carps, penaeid shrimp species, and giant freshwater prawns used in culture were successfully hatched under controlled conditions in hatcheries.

Since the 1970s, the aquaculture sector has seen continuous development, with an increasing number of species being brought into culture. The industry has expanded both in terms of geographical coverage and production quantity. Operators in the industry have adopted a new trend of selecting species that offer the highest profitability for culture, particularly emphasizing high-value species with strong export demand such as penaeid shrimps, high-value finfish (seabass/ groupers), seaweeds, and related species.

As demand and market value for selected species continue to rise, the industry has embraced high-technology methods and intensified operations. Competition for major markets and the maintenance of product quality standards have become significant concerns within the industry.

History of Aquaculture in India

Evidence of fish utilization in India dates back to three millennia BC, with fish remains and cut marks discovered during excavations at Mohenjodero and Harappa of the Indus Valley Civilization (2500 BC – 1500 BC). In 300 B.C., Kautilya described in his "Artha Shastra" how fish could be poisoned in tanks during warfare. King Someswara, son of King Vikramaditya VI, documented India's common sport fishes in his book "Manasoltara," compiled in 1127 AD, thus providing early classification of marine and freshwater forms.

During British rule in India, sport fisheries flourished with the introduction of trouts in the hill streams of Nilgris, Kashmir, and Kulu valley. Fisheries departments were established to encourage the culture of food fishes and sport fishes, with the first scientifically designed fish farm constructed by the Madras fisheries department at Sunkesula in Krishna district (now Andhra Pradesh) in 1911. Fisheries departments were further established for aquaculture development in West Bengal, Punjab, Uttar Pradesh, Andhra Pradesh, and Karnataka during 1908-1947.

Previously, fry were collected from wild waters for aquaculture. Technological breakthroughs in induced spawning of cultivable species occurred from 1700 to 1900, addressing the urgent need for seeds to support the expanding aquaculture industry. Indian scientists achieved the first success in induced breeding of Indian major carp through hypophysation in 1957, followed by Chinese success with Chinese carp in 1958. Shrimp species and giant freshwater prawns used in culture were also successfully hatched under control in hatcheries.

The establishment of the Pond Culture Division at Cuttack in 1949 under the name of the Central Inland Fisheries Research Institute (CIFRI), West Bengal, marked a significant milestone in the development of freshwater aquaculture in India. Brackishwater farming, mainly confined to bheries (manmade impoundments in coastal wetlands) of West Bengal and pokkali (salt-resistant deepwater paddy) fields along the Kerala coast, gained recognition with the initiation of the All India Coordinated Research Project (AICRP) on 'Brackishwater Fish Farming' by ICAR in 1973.

While the project developed technologies for fish and shrimp farming, scientific and commercial culture currently focuses predominantly on shrimp farming. The earliest attempt at mariculture in India was made at the Mandapam center of CMFRI in 1958–1959 with the culture of milkfish (*Chanos chanos*). CMFRI has since developed various technologies for species such as oysters, mussels, clams, shrimp, and finfish. Additionally, CMFRI initiated a successful pearl culture program in 1972, contributing to the development of pearl production in Indian pearl oysters.

Current Global Aquaculture Scenario

According to SOFIA 2022 data, Global fisheries and Aquaculture Production reached around 214 Mt in 2020, although it has suffered from severe Covid-19 impacts. This total comprises 178 Mt of Aquatic animals and 36 Mt of Algae. In 2020, global production of aquatic animals slightly decreased to 178 million tonnes from the all-time high of 179 million tonnes recorded in 2018. Out of this production, a total of 157 Mt were used for direct human consumption. In 2022, Percapita consumption of Aquatic foods is slightly decreased to 20.2 Kg/ Annum due to the COVID-19 pandemic. The fisheries and Aquaculture sector not only produce food but also provides livelihood to the 59 million people over 90% of these in small-scale fisheries. According to SOFIA 2022 data, China is in the 1st Position in Aquatic animal Production all over the world. India is in 2nd Position in World Aquatic Animal Production. In inland aquaculture production of finfish, China is in 1st Position and India is in 2nd Position. In marine and coastal aquaculture production of finfish, China is in 1st Position and India is in 13th position.

Current Indian Aquaculture Scenario

India holds significant importance in the global aquaculture landscape. According to the Department of Fisheries, India's total fish production for the 2019-2020 period stood at 14.16 million metric tonnes (MMT), comprising 3.72 MMT from marine sources and 10.43 MMT from inland sources. In 2020, the annual growth rate in fish production reached 4.35%. Gujarat emerged as the leading position in marine fish production, achieving a maximum output of 7.01 lakh tonnes, while Andhra Pradesh led in inland fish production, reaching a peak of 36.1 lakh tonnes. The fisheries sector witnessed a total expenditure of ⍰ 64,025.86 lakh in 2020 towards its development. The country boasted 1548 fish landing centers, including 7 major fishing harbours and 62 minor fishing harbours. Tripura, Kerala, Manipur, Odisha, and Assam were among the top five states in fish consumption during 2020.

Aquaculture Versus Agriculture

Agriculture

India's agricultural legacy traces back to around 9000 BC, marked by the early cultivation of crops and domestication of animals, fostering settled lifestyles and advancements in agricultural practices. Benefiting from a dual monsoon system, India has historically yielded two harvests annually, contributing to its agricultural prowess. Presently, agriculture ranks as India's primary economic sector, constituting 16.6% of GDP in 2007 and engaging 52% of the workforce. Noteworthy crops include paddy rice, wheat, buffalo milk, cow milk, and sugarcane, with India being a global leader in several agricultural commodities.

Established in 1905, the Indian Agricultural Research Institute (IARI) played a pivotal role in catalyzing the "Indian Green Revolution" of the 1970s. The Indian Agricultural Statistics Research Institute specializes in innovative techniques for agricultural experimentation and statistical analysis.

Despite significant strides, challenges such as erratic monsoons, escalating input expenses, labor scarcities, and dwindling agricultural profits persist, compounded by the encroachment of agricultural lands for real estate purposes. Addressing these constraints necessitates enhancing productivity while curbing input costs, given the global reliance on agriculture for food security.

Aquaculture

Aquaculture, often termed as "underwater Agriculture," has witnessed notable growth, mainly driven by the depletion of wild fish and shellfish stocks. In 2008, China led global aquaculture production, contributing 62 percent of fish, crustaceans, and mollusks output, totaling 32.7 million tonnes. Other major producers included India (3.5 million tonnes), Vietnam (2.5 million tonnes), Indonesia (1.7 million tonnes), Thailand (1.4 million tonnes), and Bangladesh (1 million tonnes).

Carp species dominate global aquaculture, constituting 39 percent of total production. Shellfish (like oysters, clams, mussels, and scallops), freshwater fish (including tilapias), shrimps, prawns, and salmon are also significant. In terms of value, white leg shrimp led in 2008, generating USD 9 billion, followed by Atlantic salmon (USD 7.2 billion), grass carp, and silver carp (both USD 4.8 billion).

India, with substantial marine resources and biodiversity, hosts over 10 percent of global fish species. While India has made strides in freshwater fish and shrimp aquaculture, development in mollusk culture, seaweed culture, and certain marine fish like seabass remains below commercial scale. There's a noticeable production gap between China, India, and other key Asian aquaculture nations. Leveraging India's extensive coastline for aquaculture could significantly boost production.

Agriculture	Aquaculture
Cultivation of Crops	Culture of aquatic organisms
Agriculture has begun by 9000 B.C	Aquaculture has begun during 2000-1000 BC
India ranks second position in the field of Agriculture	India ranks 2nd Position in the field of Aquaculture
India is among the world's leading producers of paddy rice, wheat, buffalo milk, cow milk and sugar cane. It is also the second largest producer of wheat and cow milk.	India is in the leading position in culturing of carps and other shellfish species such as White leg shrimp (Penaeus vannamei)
The Indian Agricultural Statistical Research Institute (IASRI) develops new techniques for the design of agricultural experiments, analyses data in agriculture, and specializes in statistical techniques for animal and plant breeding.	Central Institute of Freshwater Aquaculture (CIFA), Central Marine Fisheries research institute (CMFRI), Central Institute of Brackishwater Aquaculture (CIBA) and some of the fisheries research institutes are developed innovative culture techniques in Aquaculture.

Advantages of Aquaculture

- Aquaculture serves as a vital source of high-quality protein and healthy oils, ensuring a sustainable future for fish production.
- The low cost of fish production in aquaculture allows for affordable pricing, making it accessible even to economically disadvantaged communities.
- Cultured fishes are free from pollutants, ensuring their safety and quality, making them a reliable source of food for the growing global population.
- Additionally, aquaculture fosters employment opportunities, contributing to economic growth and stability.

Disadvantages of Aquaculture

- Aquaculture infrastructure development can impact local ecosystems, including wetlands and mangroves, while untreated effluent discharge with heavy organic load can adversely affect local ecosystems. Farming exotic species could introduce new pathogens to the environment and increase the risk of disease and parasite transfer from captive stock to the wild.
- In the current era of food insecurity, aquaculture has great potential to feed not only the increasing human population but also to provide feed ingredients for high-value domesticated animals. Global developments and the strategic importance of aquaculture in terms of food security contribute to its promising future. Aquaculture has become a modern, dynamic industry that produces safe, high-value, and high-quality products, with sustainable environmental practices. In recent years, aquaculture has led to substantial socioeconomic benefits, including increased nutrition, income, employment, foreign exchange, and the development of previously underutilized land and water resources.
- The world faces the challenge of a "food gap," which is expected to more than double in the developing world over the next 25 years, leading to increased dependence on food imports from developed countries. To address the food gap, improvements in sustainable agriculture and aquaculture production are essential.

Chapter-3

Systems of Aquaculture

Aquaculture encompasses various classifications based on different aspects and practices involved. Some fundamental classifications include:

Salinity-based Categorization

- **Freshwater farming:** Involves cultivating aquatic animals and plants in zero saline water.
- **Brackishwater farming:** Refers to the cultivation in water with a salinity less than 30ppt, found in estuaries, backwaters, creeks, and mangrove waterways.
- **Marinewater farming:** Involves farming aquatic organisms in seawater, also known as mariculture.

Intensity-Based Classification

Extensive Fish Farming System

- Characterized as the least managed form of farming, utilizing large ponds (1 to 5 hectares) without supplemental feeding or fertilization.
- Stocking density is relatively low, yielding 500 to 2000 kilograms per hectare.

Semi-intensive Fish Farming System

- Utilizes smaller ponds (0.5 to 1 hectare) with moderate stocking density and natural food development, sometimes supplemented with feed.

- Yields range from 2500 to 10,000 Kg for fish and 1500 to 2500 Kg for shrimp per hectare.

Intensive Fish Farming System

- A well-managed approach aimed at achieving maximum production from minimal water resources.
- Involves small ponds, tanks, or raceways with high stocking density, complete reliance on formulated feed, and rigorous management practices.
- Yields can range from 15 to 100 tons per hectare or more, with higher investment costs but correspondingly high yields compared to other systems.

Species-based Classification

- **Monoculture:** Involves rearing only one species in a culture system, such as trout, tilapia, catfishes, carps, or shrimp.
- **Polyculture:** Involves farming two or more different fish species together, often resulting in higher yields due to complementary feeding habits and behaviors.

Enclosure-based Classification

- **Pond culture:** Common method involving enclosed water areas constructed with dikes or bunds, typically filled by rain, canal water, or manmade bores.
- **Cage culture:** Rearing fish in enclosed volumes of water, suitable for areas where draining is not feasible, with advantages in stock management and monitoring but susceptible to external water quality issues and predation.

Advantages of Cage Culture

- Implement straightforward technical procedures for setting up or expanding farms
- Reduce capital expenditures compared to land-based farms
- Simplify stock management and monitoring compared to pond culture

Disadvantages of Cage Culture:

- Vulnerable to external water quality problems *eg.* Algal blooms, low
- oxygen
- Stock is more vulnerable to predators such as birds
- Growth rates are significantly influenced by ambient water
- parameters

Pen culture: Raising fish in enclosed water volumes, allowing for free water circulation while providing protection from predators, offering benefits such as increased stocking density and ease of harvest.

Advantages of Pen Culture

- **Efficient space utilization:** Pen culture allows for higher stocking density compared to pond systems.
- **Predator protection:** Enclosures offer safety from predators, especially in smaller pens.
- **Versatility for diverse species:** Pen culture suits various species due to ample space and natural water systems.
- **Controlled harvesting:** Harvesting in large pens is more manageable than in natural waters.
- **Size flexibility and cost-effectiveness:** Pens can be constructed larger and more affordably than cages.
- **Natural food availability and bottom exchange:** Cultured organisms benefit from natural food and material exchange with the pen bottom.

Disadvantages of Pen Culture

- **Oxygen and water flow demands:** Pen culture requires sufficient oxygen and water flow to support fish.
- **Reliance on artificial feed:** Fish in pens heavily depend on artificial feed.
- **Food loss:** Some feed may go uneaten and drift away, albeit less than in floating cages.
- **Pollution risk:** Intensive culture can lead to pollution from accumulated excrement and substances like ammonia if not managed.
- **Disease spread:** High stocking density can facilitate rapid disease transmission, causing high mortality.
- **Theft vulnerability:** Enclosed pens may be susceptible to theft.
- Water use conflicts: Pen culture may clash with other water uses like irrigation and recreation.

Zerowater Exchange Systems in Aquaculture

1. Recirculatory Aquaculture System

A Recirculating Aquaculture System (RAS) is a method of aquaculture that involves treating and reusing water, with less than 10% of the total water volume replaced daily. Fish species such as channel catfish and rainbow trout have been effectively cultured using RAS.

One notable strategy for increasing production while minimizing waste is the implementation of recirculating aquaculture systems (RAS). These systems are engineered to collect and eliminate waste products, uneaten feed, and bacteria from the fish tank, allowing water to be recycled within the system. RASs can be configured for indoor or outdoor tanks or ponds, with variations in design and construction. However, most RAS perform key functions in wastewater treatment:

- Removal of solid waste.
- Removal or detoxification of nutrients.
- Removal of carbon dioxide.
- Supplementation of dissolved oxygen.
- Sterilization of bacteria and pathogens.

2. Raceways

Running water fish culture typically occurs in earthen ponds sized around 100 m^3 (10m x 10m x 1m) at high altitudes. Suitable fish species for this practice include trout, mahseer, and common carp. It's an intensive system requiring high stocking density and protein-rich feeds. Raceways, shallow tanks relying on high water flow, sustain aquatic life. Flow-through systems pass water through once, discharge it after waste treatment, and do not recirculate. Inflowing water temperature should match the species' tolerance. Oxygen is provided and removed by the fish as water progresses. Dissolved metabolites are carried out in the effluent, and waste is managed through settling or filtration. Nutritionally complete diets are essential. Flow-through systems, common in the U.S., are used for various coldwater species like trout and salmonids, as well as warm water species like catfish and tilapia. Raceways consist of tanks or series of rectangular tanks with water flowing along the long axis. Linear-pass raceways exhibit a water quality gradient from inflow to outflow, while circular units offer more uniform conditions. Raceways provide higher production per unit of space and better fish observation, feeding efficiency, disease detection, growth monitoring, and inventory estimates compared to ponds.

Based on Integration

Integration With other Farming Practices:

Integration with agriculture, paddy farming, horticulture, mushroom cultivation, sericulture, and livestock farming allows for diverse and complementary practices, enhancing productivity and sustainability. Each integration type offers unique advantages and considerations for farmers, contributing to the overall efficiency and profitability of aquaculture operations.

Paddy Fish Integrated Farming

In regions where paddy fields are submerged for 3 to 8 months annually, integrating fish farming with rice cultivation can supplement fish production. While historically practiced in various countries like Japan, Malaysia, Italy, China, and India, the compatibility of fish culture with modern high-yielding paddy varieties is diminished due to widespread use of aquatic-life-toxic pesticides and chemicals. However, in some northeastern states of India, this traditional method persists. Leveraging flooded paddy fields, this integrated approach can yield over 300 kg of fish per hectare alongside rice production. Commonly cultivated fish species in this system include carps like Cyprinus carpio IMC, while rice varieties with robust root systems, such as Tulsi, Panidhan, CR260 77, ADT 6, ADT7, Rajarajan, Pattambi 15, and Pattambi 16, are preferred.

Horticulture Fish Integrated Farming

The horticulture-fish farming system integrates the cultivation of fruits, vegetables, and flowers along the pond embankments. Fruits and vegetables, rich in essential nutrients, are recommended by the Indian Council of Medical Research for daily consumption. Utilizing the inner and outer dykes of the pond and adjacent areas, plant selection is crucial for system success, favoring dwarf, seasonal, evergreen, and economically viable species with minimal shade. Fruit options include Mango, Banana, Papaya, Coconut, Lime, while vegetables like Brinjal, Tomato, Cucumber, Gourds, Chilli, Carrot, Radish, Turnip, Spinach, Peas, Cabbage, Cauliflower, and Ladies finger can be grown year-round. Additionally, flower cultivation with species such as Rose, Jasmine, Gladiolus, Marigold, and Chrysanthemum adds aesthetic value and supplemental income. This integrated system typically yields 20-25% higher returns compared to standalone aquaculture.

Mushroom Fish Integrated Farming

Cultivating mushrooms demands a high level of humidity, making their integration with aquaculture highly promising. In India, commercially cultivated mushrooms include *Agaricus bisporus*, *Volvariella* spp, and *Pleurotus*.

Seri Fish Integrated Farming

In this farming approach, silkworms are raised alongside fish. Mulberry leaves, the primary food for silkworms, are utilized by them, and their waste is directly applied to the fish pond to enhance the natural food organisms, detritus, and bacteria present.

Livestock Fish Integrated Farming

In this integrated farming system, the excrement of ducks, chicks, pigs, and cattle is directly utilized in ponds to boost plankton production, which serves as food for fish or is consumed directly by them.

Cattle Fish Integrated Farming

For a 1-hectare pond, 5-6 cows can supply sufficient manure. An average healthy cow excretes between 4,000-5,000 kg of dung and 3,500-4,000 liters of urine annually. Fish production in this setup ranges from 3,000-4,000 kg per hectare per year, while milk production amounts to 9,000 liters per year.

Duck Fish Integrated Farming

Cultivating fish alongside livestock, particularly ducks, offers several advantages. Ducks naturally consume various aquatic organisms such as insects, crustaceans, and mollusks, which may not be economically viable otherwise. Moreover, duck droppings serve as both food and fertilizer for the ponds. As ducks forage in the pond, they release nutrients from the soil and enhance oxygen mixing in the water, thereby boosting biological productivity and facilitating fish growth. Ducks, which typically prefer aquatic habitats, require minimal housing. High-quality duck breeds like India Runners are often used in this integrated system. Cultivating 200-300 ducks alongside fish can provide ample fertilizer for the pond. This integrated approach can yield 3,500-4,000 kg of fish, 18,000 eggs, and 500 kg of duck meat per hectare annually.

Pig Fish Integrated Farming

Combining fish farming with pig rearing offers a cost-effective opportunity and an additional income stream for fish farmers. In India, Hampshire and Land Race are among the commonly cultured pig breeds for this purpose. These breeds reach slaughterhouse maturity (60-70 kg) within six months and typically produce 6-8 piglets. Fish reach marketable size within a year, allowing for two cycles of pig rearing annually. Pig manure, rich in nutrients comparable to cow dung, serves as valuable fertilizer. A fully grown pig can produce 500-600 kg of manure annually. Approximately 30-40 pigs per hectare are sufficient for adequate fertilization. This integrated system can yield approximately 6,000- 7,000 kg of fish per hectare per year, along with 3,600-5,000 kg of pig meat.

Poultry Fish Integrated Farming

Chicken droppings, abundant in phosphorus and nitrogen, serve as valuable fertilizer. With 25,000 chicks reared per hectare of fish pond, poultry farming can yield between 90,000 to 100,000 eggs and 2,500 kg of meat. Fish production in this integrated system ranges from 3,000 to 4,500 kg. The productivity of poultry in terms of meat and egg production depends on various factors such as breed, variety, feeding practices, healthcare, and management. Housing 250 birds per pen on litter-covered floors, integrated fish cum poultry farming can generate approximately 4,000-4,500 kg of fish, over 70,000 eggs, and 1,250 kg of chicken meat per hectare of pond annually.

Chapter-4

Pond Culture

This method of fish culture involves maintaining water in enclosed areas through artificially constructed ponds, where various aquatic animals like finfish and shellfish are raised. Ponds can be filled with canal water, rainwater, borewell water, or other water sources, with proper consideration given to site selection based on climate, topography, water availability, and soil quality.

There are two primary types of ponds based on these factors:

Watershed Pond Systems

- These ponds rely entirely on watershed runoff to fill and maintain water levels, although groundwater and surface water can supplement the supply.
- Factors such as soil type, topography, and water quality are crucial considerations.
- Soil type, particularly mud or mud-silt soils, is important for preventing leakage.
- Topography dictates pond size and shape, with a maximum depth of 10 feet recommended.
- Management is more challenging due to erratic water supplies and uneven pond bottoms.

Levee Pond Systems

- Constructed in flat land areas where watershed runoff is insufficient, these ponds are typically filled with groundwater.

- Soil characteristics, groundwater availability, and topography are essential factors for site suitability.
- Adequate clay content (at least 20%) in the soil prevents leakage.
- Rectangular pond shapes are preferred for simplicity and ease of management.
- Water quality considerations are vital for successful pond construction.
- Levee ponds are more suitable for fish growth and harvesting compared to watershed ponds.

Hybrid Pond

These ponds represent a hybrid approach, blending features of both embankment and watershed ponds. They typically feature embankments on two or three sides, enclosing a relatively small, shallow drainage basin. Hybrid watershed-embankment ponds are constructed in areas with gently rolling topography, where neither pure embankment nor watershed ponds are ideally suited.

Ponds in Aquaculture Farm

Nursery Pond

After hatching, spawn develop into larvae over three days. These larvae are nurtured until they reach the fry stage (2-3 cm length), typically taking 11 to 30 days, in nursery ponds. Each fry pond ideally spans 0.01 to 0.05 ha, with numbers adjusted based on production targets, and maintains a water depth of 1-1.5 m. The nursery pond is typically rectangular in shape with a flat, even bottom to facilitate straightforward netting operations.

Rearing Pond

In a similar manner, fry are raised to the fingerling stage (10-15 cm length) within rearing tanks, with the duration typically spanning 2-3 months. Fingerling ponds range in size from 0.05 ha to 0.1 ha, with water depths of 1.5 - 2 m, and their quantity aligns with the target production of fingerlings. These ponds are rectangular in shape with a flat, even bottom to facilitate convenient netting operations.

Stocking Pond

The stocking ponds, ranging from 1 ha to 2 ha in size, are considered optimal for intensive food fish culture, with water depths of 2.5 - 3 meters. Fingerlings measuring 10-15 cm in length are introduced into these ponds until they reach table size or marketable proportions. The number of stocking ponds varies depending on the target fish production, and the rearing period typically spans 8-10 months. These ponds may also serve as breeding or brooder raising ponds as needed, with flexibility regarding their size.

Bio Pond or Treatment Pond

In contemporary fish farming practices, alongside traditional fish ponds, a specialized type known as the Bio pond has emerged. Serving as a large settling tank, it functions to biologically purify the water used within the fish farm ponds. Additionally, it can be utilized as a stocking pond when necessary. Typically, this type of pond covers approximately 7-10% of the total productive area of a fish farm.

Ponds in Aquaculture farm

Types of Pond	Culturing stage	Size of fish culturing	Size of Pond	Depth	Culture period	Total farm area
Nursery Pond	Spawn (3 days old to fry)	2-3 Cm	0.01-0.05 ha	1-1.5 m	30 days	5%
Rearing Pond	Fry to Fingerling	10-15 Cm	0.05-0.1 ha	1.5-2 m	2-3 months	20%
Stocking Pond	Fingerling to Marketable size	25-40 cm	1-2 ha	2.5-3 m	8-10 months	70%

Chapter-5

Cage Culture

Cage culture, a method of fish farming, involves raising fish from juvenile to commercial size in a confined water volume, enclosed on all sides, including the bottom, while allowing water circulation. It's particularly suited for areas like lakes, reservoirs, and riverine pools where draining isn't possible. Widely practiced in countries such as Japan, South Korea, China, the Philippines, Thailand, Malaysia, Germany, Norway, and the USA, virtually all cultivable fish species, like carps, tilapia, trout, and catfish, can be successfully raised in cages, depending on socioeconomic, ecological, and technical factors. Cost-effective and simplifying fish health monitoring, cage culture enables complete harvesting without fish escaping. It's typically practiced in water bodies with minimum depths of 5 meters in reservoirs, 3 meters in wetlands, and 2 meters in canals. Commonly cultured freshwater fish and shellfish suitable for cage culture include Indian Major Carps, Chinese Carps, air-breathing catfishes, climbing perch, tilapia, snakeheads, and freshwater prawns.

Site Selection for Cage Culture

For cage culture, it's best to select sites with shallow, weed-free waters and moderate flow to support optimalfish growth. The chosen location should have sufficient water circulation, minimal pollution, and easy accessibility while being situated away from the shore to prevent poaching and crab disruptions. While cage culture can also be conducted in swampy areas, it thrives in flowing waters with a slow current of 1-9 m/minute. It's essential to understand that cage culture isn't compatible with open pond culture methods.

Construction of Cages

To construct cages, essential materials include frames, netting materials, lids, floats, nuts, bolts, sinkers, and anchors. The frame can be made of various materials like wood, bamboo, PVC, fiberglass, steel, zinc, or aluminum. Netting materials such as polyester, polyethylene, polypropylene, or netlon are commonly used. Floats, sinkers, and anchors are crucial for maintaining the cage's shape and position.

The choice of materials depends on the type of culture, whether for fry or table fish rearing. For example, bamboo interspersed with wooden planks is common in Indonesia, Vietnam, Thailand, and Kampuchia. Thick polythene fibers are used in Japan, and an aluminum frame with nylon webbing is used in the USSR and West Germany.

In India, small cages are made with mats of locally available plant materials like palm leaves, Cyperus stem, Phragmites stem, and split bamboo, typically 1-2 m^2 in size. These cages are installed in the water body with bamboo supports at the four corners and the bottom. While materials other than bamboo mats decay within a year, split bamboo cages can last for over a year. Circular cages made with thick bamboo poles tied with nylon twine can have a durability of over 3 years.

Feeding of Fish in Cages

Given the high stocking density and the confinement of fish in cages, natural food supply from flowing water is inadequate, necessitating regular provision of artificial feed. Typically, feed is provided in the form of pellets, comprising approximately 70% animal and 30% plant material. Raw materials for these pellets include rice bran, soybean, oil cake, meat, liver, blood, and fish. The amount of food supplied varies depending on the species and stocking density within each cage, with feeding carefully regulated to ensure complete consumption before being washed away by water currents.

Pangasius, a robust and carnivorous fish species, feeds on gastropods, mussels, and insects, making it well-suited for culture in cages and pens constructed from wood or bamboo. Stocking rates typically range from 1000 fingerlings per medium-sized cage to 150-3000 in larger ones. Artificial feed, comprising chopped trash fish, banana, and rice bran, is provided at a rate of 4-6% of the fish's weight per day. Under optimal conditions, Pangasius can reach a size of 1-4 kg within 8-10 months.

Types of Cages

Cages are available in circular, cubic, and basket-like configurations, with their shape having minimal impact on the yield rate. They can either float on the surface, be submerged, or rest on the bottom of the water body. Floating cages are often the preferred choice for Indian conditions, with various experiments conducted in such enclosures for seed rearing, grow-out, nutrition, and biomonitoring purposes. The size of the cage is determined by the type of culture operation and the support facilities available. While larger cages pose handling challenges, smaller cages are easier to manage and entail lower risk of losses. Typically, nursery cages are of the floating type, whereas ground cages can either float or be immersed, depending on the species being cultured.

Advantages of Cage Culture

Take advantage of pre-existing water bodies

- ❑ Implement straightforward technical procedures for setting up or expanding farms
- ❑ Reduce capital expenditures compared to land-based farms
- ❑ Simplify stock management and monitoring compared to pond culture
- ❑ Easy harvesting of stock
- ❑ Disease treatment can be effected more accurately and economically.

Disadvantages of Cage Culture

- ❑ External water quality issues, such as algal blooms and low oxygen levels, pose risks to fish stocks
- ❑ Fish stocks face increased vulnerability to predators, including water rats and birds
- ❑ Growth rates are significantly influenced by ambient water temperatures
- ❑ Fouling of the cage netting is also a common Problem
- ❑ Poaching and vandalism can be devastating problems in cage aquaculture in both public and private waters.
- ❑ Excessive current cause problems with moorings and deformation of cage structure and integrity

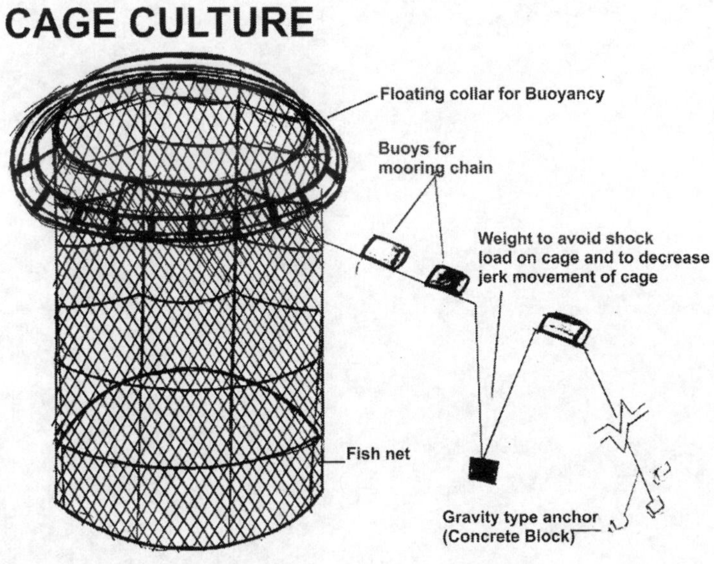

Chapter-6

Pen Culture

Pen culture involves confining fish in a designated water volume, enclosed on all sides except the bottom, enabling water circulation from at least one side. Combining aspects of pond and cage culture, this method typically employs net or wooden materials to create pens along shallow regions of lakes and reservoirs' shores and banks. These pens utilize the lake's bottom, which harbors benthic fauna as fish food. Polyculture, akin to pond culture, can be practiced in these pens, which maintain a water environment with ample dissolved oxygen due to free water exchange with the surrounding body. This method finds extensive use in Japan, Peru, and the Philippines, where, for example, milkfish fingerlings are reared in pens in Laguna de Bay until reaching marketable size (200 g or more). Typically constructed using split bamboo or netting, these pens house fish from post-monsoon to late autumn, with a culture period spanning 6 to 8 months. Notably, major carp seed rearing has been an annual practice since 1982 in the Tungabhadra reservoir in Karnataka.

Selection of Sites for Pen Culture

When selecting sites for pen culture, optimal locations feature minimal tidal fluctuation, protection from strong winds, and a minimum depth of 1 meter during low water levels. The most desirable spots are on the side sheltered from prevailing winds, with moderate current flow, especially in areas where currents regularly circulate. Stable pH levels and slight fluctuations in water quality are preferred, while turbid or polluted water should be steered clear of.

Construction of Pens

Construction of fish pens using bamboo screens does not require specialized shaping or rounding of the bamboo poles. The process involves soaking the bamboo poles in water for two weeks followed by a week of drying. During this period, the poles are prepared and staked at the selected site to define the desired size and shape of the fish pen. Once staked, bamboo splits are tightly woven into rolls approximately five meters long. These woven splits are then stretched between poles, interwoven or placed inside or outside close to the poles from bottom to top. They are secured to each pole using rubber, with additional rubber slices placed around each pole, top and bottom, to prevent wear from wave action.

Supplementary Feeding

Fish pens with high stocking densities, ranging from 10 to 20 fish per square meter, typically require regular feeding. Feeding rates usually range from 4 to 10% of the total body weight of the stock, administered at least once every three weeks or divided into daily feedings. The amount of feed given is adjusted based on the condition of the cultured fish, which is monitored through monthly sampling.

Management of Fish Pens

Managing a fish pen is more labor-intensive and challenging compared to afish farm due to increased risks. Fingerlings are particularly susceptible to escaping if even a single bamboo split breaks or if there's a small tear in thenet.

Types of Pen Enclosures in Aquaculture

- ❑ Isolated enclosures, fully surrounded by a net structure, positioned in the center of a bay without any foreshore
- ❑ Shore enclosures, extending a portion of the foreshore into deep water, enclosed by a net structure
- ❑ Bay or lock enclosures, featuring an embankment equipped with sluices or anet structure only at the entrance

Types of Pens in Aquaculture

Rigid Pens:

- ❑ **Embanked Pens:** Constructed with concrete, stone, or earth embankments featuring sluices for water circulation.
- ❑ **Net Enclosures:** Comprising frameworks or scaffolding with walls made of mesh netting.

Flexible Pens

Components include sinkers, chain-link footrope, headrope, flotation buoys, netting wall (vertical and horizontal floating net), strengthening lattice, catwalks, and jetties.

Outer Barrier Nets

Selection Criteria for Netting Material

The choice of netting material is based on several factors, including:

- Reasonable life expectancy
- Resistance to fouling and corrosion
- Ease of handling and cleaning

Physical Properties of Netting Materials

- Resistance to environmental and climatic factors
- Relative efficiencies, such as curved corners versus square corners

Additional Features

- Provision of collars to prevent jumping fish
- Footings and surface covers to deter predators
- Construction of catwalks, feeding platforms, and boat locks.

Advantages

- **Efficient space utilization:** Pen culture allows for higher stocking density compared to pond systems.
- **Predator protection:** Enclosures offer safety from predators, especially in smaller pens.
- **Versatility for diverse species:** Pen culture suits various species due to ample space and natural water systems.
- **Controlled harvesting:** Harvesting in large pens is more manageable than in natural waters.
- **Size flexibility and cost-effectiveness:** Pens can be constructed larger and more affordably than cages.
- **Natural food availability and bottom exchange:** Cultured organisms benefit from natural food and material exchange with the pen bottom.

Disadvantages

- **Oxygen and water flow demands:** Pen culture requires sufficient oxygen and water flow to support fish.
- **Reliance on artificial feed:** Fish in pens heavily depend on artificial feed.
- **Food loss:** Some feed may go uneaten and drift away, albeit less than in floating cages.
- **Pollution risk:** Intensive culture can lead to pollution from accumulated excrement and substances like ammonia if not managed.
- **Disease spread:** High stocking density can facilitate rapid disease transmission, causing high mortality.

- **Theft vulnerability:** Enclosed pens may be susceptible to theft.
- **Water use conflicts:** Pen culture may clash with other water uses like irrigation and recreation.

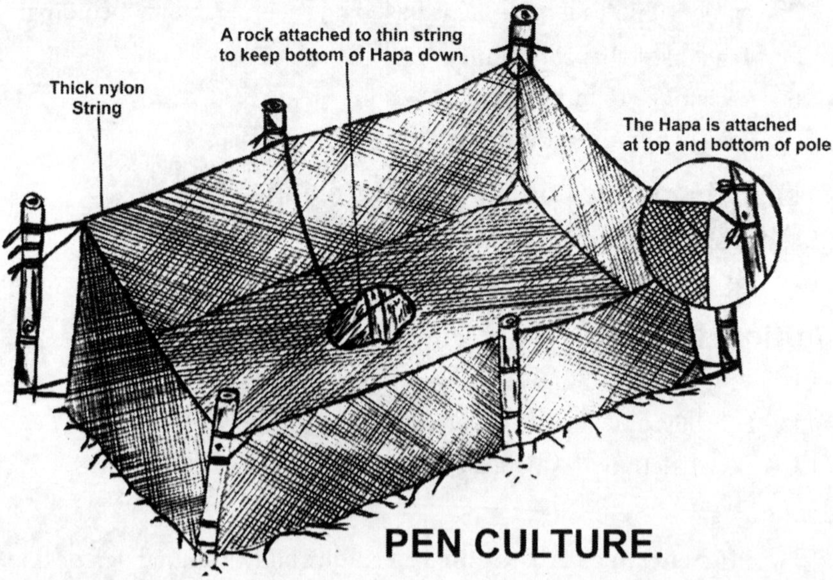

PEN CULTURE.

Chapter-7

Integrated Fish Farming

Integrated fish farming operates on the principle that no resource should go to waste; instead, waste should be viewed as a potential valuable material for other products (FAO, 1977). The fundamental principles of integrated farming emphasize leveraging the synergistic effects of interconnected farm activities and the conservation, as well as the full utilization, of farm wastes.

Fish farming can be integrated with various agricultural practices to enhance productivity and sustainability. These integrated systems include rice-fish integration, horticulture-fish system, mushroom-fish system, and seri-fish system.

Paddy Cum Fish Culture

The water-logged rice fields, submerged for 3-8 months annually, naturally harbor a small population of fish. This natural occurrence likely led to the intentional introduction of fish stocks for subsequent harvesting. Traditional methods, such as using "gamcha" or "dhoti" cloth to trap prawns and fish in fallow paddy fields, have been practiced for generations in India.

Objectives of Paddy Fish Culture

- Paddy field aquaculture offers farmers an opportunity to supplement their income.
- In regions where rice and fish are dietary staples, paddy-field aquaculture ensures access to vital food sources for the local population.
- Growing both rice and fish concurrently or interchangeably in the same water body minimizes additional expenses, especially in terms of management and labor.

This practice offers employment opportunities to farmers and agricultural workers during the off-season.

Combining paddy and fish farming mutually benefits both. Fish cultivation contributes to improved rice production by effectively controlling unwanted weeds, mollusks, harmful insects, and their larvae.

Fishes Suitable for Paddy Cum Fish Culture

Not all fish species are suitable for cultivation in paddy fields due to the unique ecological conditions they offer, such as shallow, turbid water and high temperatures.

Paddy-field aquaculture typically involves three main types of practices:

- ❑ Simultaneous or combined cultivation.
- ❑ Alternating or sequential farming, also known as paddy-fish rotation.
- ❑ Relay farming.

Simultaneous or Combined Cultivation

In this farming approach, both paddy and fish are cultivated concurrently, and the harvest of both crops occurs at the conclusion of the rice-growing season.

Research indicates that various techniques and stocking rates have been utilized in different countries for paddy-field aquaculture. Typically, fish stocking begins no sooner than 5 days after the transplantation of paddy seedlings, allowing the seedlings adequate time to establish proper root systems. After paddy transplantation, fry stocking should be conducted after 10 days, whereas fingerlings can be stocked after 3 weeks.

Alternating or Sequential Farming, Also Known as Paddy-Fish Rotation.

In this basic approach, flooded paddy-fields are utilized to cultivate one or more crops of fish or shrimp following the rice harvest.

In the rotation system, preparation of the paddy-field for fish cultivation occurs after the rice harvest. To maintain the necessary water depth, the bunds surrounding the field are elevated. Following rice harvesting, the stubbles are left in place, submerged beneath the water, providing a substrate for the growth of fish food organisms. As these stubbles decompose, they enrich the water, enhancing productivity. After fish harvesting, the residual matter remains in the soil, serving as fertilizer for the subsequent paddy crop.

Relay Farming

This approach is more intricate, facilitating an extended period of fish cultivation. It entails relocating the fish stock to specially prepared ditches, channels, or pools at the time of paddy harvest, and subsequently reintroducing them to the field for further growth. This method allows for a better growth rate than what can be achieved within the duration of a single paddy crop.

For such cultivation, the farm needs additional holding or rearing facilities during the period between paddy harvesting and planting new fish seedlings. Carps or shrimps can be raised to the desired size through adequate feeding. However, this method requires greater inputs, including labor.

Rice Fish Integrated Farming

In rice-fish integrated farming, certain rice varieties with robust root systems are combined with fish species like Common carp, Tilapia, and Murrells, optimizing flood resilience and fish cultivation.

Horticulture-fish integrated farming utilizes dykes and pond surroundings for planting dwarf coconut, mango, banana, pineapple, ginger, turmeric, and chilly. Water exchange from fish ponds enriches plants with organic nutrients, while crop residues can be recycled into the pond to support species like grass carp.

Mushroom-fish integrated farming capitalizes on the humidity required for mushroom cultivation, offering opportunities for combined aquaculture.

Seri-fish integrated farming involves raising silk worms alongside fish, with mulberry leaves serving as primary silk worm feed and their waste enhancing natural food organisms in fishponds.

Livestock integrated fish farming, in various systems like cattle-fish, pig-fish, poultry-fish, duck-fish, goat-fish, and rabbit-fish integrate animal excreta into ponds to boost plankton production, reducing the need for chemical fertilizers and supplementary feeds.

Cattle-fish integrated farming utilizes cow dung as a nutrient source, while pig-fish integrated farming relies on pig manure for fertilization. Poultry-fish integrated farming utilizes chicken manure for pond fertilization, with chickens reared above the pond area.

Duck-fish integrated farming Cultivating fish alongside livestock, particularly ducks, offers several advantages. Ducks naturally consume various aquatic organisms such as insects, crustaceans, and mollusks, which may not be economically viable otherwise. Moreover, duck droppings serve as both food and fertilizer for the ponds. As ducks forage in the pond, they release nutrients from the soil and enhance oxygen mixing in the water, thereby boosting biological productivity and facilitating fish growth. Ducks, which typically prefer aquatic habitats, require minimal housing. High-quality duck breeds like India Runners are often used in this integrated system. Cultivating 200-300 ducks alongside fish can provide ample fertilizer for the pond. This integrated approach can yield 3,500-4,000 kg of fish, 18,000 eggs, and 500 kg of duck meat per hectare annually.

Pig Fish Integrated Farming

Combining fish farming with pig rearing offers a cost-effective opportunity and an additional income stream for fish farmers. In India, Hampshire and Land Race are among the commonly cultured pig breeds for this purpose. These breeds reach slaughterhouse maturity (60-70 kg) within six months and typically produce 6-8 piglets. Fish reach marketable size within a year, allowing for two cycles of pig rearing annually. Pig manure, rich in nutrients comparable to cow dung, serves as valuable fertilizer. A fully grown pig can produce 500-600 kg of manure annually. Approximately 30-40 pigs per hectare are sufficient for adequate fertilization. This integrated system can yield approximately 6,000- 7,000 kg of fish per hectare per year, along with 3,600-5,000 kg of pig meat.

Small ruminants like goats and sheep are integrated on a smaller scale, while integrated rabbit-fish farming remains less explored.

Chapter-8

Monoculture and Polyculture Systems

Monoculture

Monoculture, exemplified by the exclusive cultivation of single fish species such as *Clarias* sp or *Oreochromis niloticus* in ponds or tanks, is prevalent worldwide. Examples include the cultivation of Common carp in East Germany and Japan, *Tilapia niloticus* in various African nations, Rainbow trout (*Salmo gairdneri*) in multiple countries, Channel catfish (*Ictalurus punctatus*) in the United States, and Catfish (*Clarius gariepinus*) in Africa.

Categories of Monoculture

- ❑ Freshwater Aquaculture
- ❑ Marinewater Aquaculture
- ❑ Brackishwater Aquaculture

Objectives of Monoculture

The objectives of monoculture include maximizing production, safeguarding against species extinction due to over-exploitation, and cultivating protein-rich fish.

Advantage of Monoculture

Monoculture facilitates ease of feeding for the fish, allowing for precise control over their size, age, and sex. It enables farmers to tailor-feed to meet the specific requirements of each fish species and offers easy monitoring of individual fish breeding performance. Additionally, there is no competition for space or feed in monoculture setups.

Disadvantages of Monoculture

In monoculture systems, the full utilization of the natural productivity of the pond is not achieved, and available space or ecological niches in the water column remain under utilized. Disease outbreaks pose a higher risk of yield loss. Additionally, there may be issues with poor water quality, cannibalism among fish, and the potential for disease outbreaks due to overstocking.

Polyculture

In polyculture, diverse species of fish with varied trophic and spatial preferences are co-cultivated in the same pond to maximize the utilization of natural food sources. The selection of species is crucial, aiming for a compatible mix that encompasses planktivorous surface/column feeders, benthic/detritivorous bottom feeders, and omnivorous to microvegetation feeding fish species.

Typical Six Carp Species in Composite Fish Culture

Species	Spatio-Trophic Habits
Silver carp	Surface feeder- Phytoplankton feeder
Bighead carp	Surface feeder- Zooplankton feeder
Catla	Surface feeder – Zooplankton feeder
Grass carp	Surface/column feeder – Macrophyte feeder
Rohu	Column feeder – Plankton and organic debris form the major diet
Mrigal	Bottom feeder- Detritivore
Common carp	Bottom feeder- Omnivore

Objectives of Polyculture

The objectives in polyculture are to achieve maximum yield or fish production, utilize all available ecological niches, and ensure that cultured fish do not disrupt the ecological balance. It's essential to select fish species that do not compete severely with each other but instead have mutually beneficial interactions, such as Grass carp which consume aquatic vegetation, converting it into fish flesh, while their excreta fertilize the pond, benefiting all other species.

Advantages of Polyculture

- Creates employment opportunities.
- Farmer has full control over intensive polyculture of fish.
- Enables higher fish cultivation and production in smaller water bodies.
- Enhances pond productivity through fish polyculture.
- Sustains water quality.
- Eliminates competition among cultured fish for food and space in the same water body.

- Maximizes utilization of spatio-trophic habitats.
- Allows for the cultivation of fish according to market demand.
- Various species combined in a polyculture setup contribute to pond health improvement.
- The defecated semi-digested food of grass carp serves as both feed for bottom-dwellers and pond fertilizer with high cellulose-digesting microbes, promoting synergistic composite fish culture.

Indian Polyculture System

In India, the initial phase of multi-species aquaculture focused on stocking Indian major carps such as rohu (*Labeo rohita*), catla (*Catla catla*), and mrigal (*Cirrhinus mrigala*). Subsequently, major carps were stocked alongside calbasu (*Labeo calbasu*), occasionally with bata (*L. bata*), and with additional species including pearl spot (*Etroplus suratensis*), gorami (*Osphronemus gorami*), and acclimatized milkfish (*Chanos chanos*), reflecting an extensive range of cultivable fish species in India.

S.No	Catla	Rohu	Mrigal	Reference
1	03	03	04	Alikuhni,1957
2	04	03	03	CIFRI,1960s
3	03	03	03	CIFRI,1960s
4	01	01	01	CIFRI,1960s

Chapter-9

Zerowater Exchange Systems in Aquaculture

Objectives

- Bacteria are fundamental to the heterotrophic, intensive, zero-water exchange shrimp culture process.
- A key component of this approach is efficient nutrient recycling in the pond, facilitated by heavy aeration. This method prevents the release of effluent, such as feces and bacteria, into natural water resources, and avoids introducing new water to prevent water-borne viruses that harm shrimp.
- Consequently, biosecurity is enhanced with this technology. Water circulation is promoted by paddlewheel aerators, which also enable production to be located inland, away from sensitive coastal environments.
- The zero-exchange system is a viable alternative to traditional pond methods of intensive aquaculture production, minimizing the associated problems.
- Under typical conditions, the Zero Water Exchange Aquaculture Production Systems (ZWEAPS) yield inexpensive and nutritious plankton feeds for aquaculture. Shrimps, fishes, and other aquaculture organisms can feed on natural biota such as phytoplankton, zooplankton, and bacteria present in culture systems.

- In intensive farming systems, microalgae and other microbial community constituents play a crucial role in nutrient recycling, decreasing anoxic zones in ponds, alleviating nutrient load in wastewater, and providing nutrition for shrimp in semi-intensive and intensive systems.
- Indoor culture enables the production of unialgal cultures while controlling contamination, but it is more expensive due to space, energy, and skilled labor requirements compared to outdoor culture. The heterotrophic condition offers advantages over photo-autotrophy in biomass production and effluent quality management, and reduces the space required for culture. High-density culture necessitates large feed quantities and extra aeration. In this scenario, waste must be effectively treated, or else culture conditions may lead to stress negatively impacting growth and productivity.
- ZWEAPS has become popular in aquaculture because it effectively controls disease outbreaks and environmental pollution.
- In discharge production systems, phytoplankton and microbial communities are crucial for recycling autochthonous nutrients accumulating within the system. These microbes, phytoplankton, and zooplankton communities are key drivers of system carrying capacity. The most promising aspect of zero water exchange culture systems is their potential to increase biosecurity and reduce feed costs and water usage, thereby enhancing the sustainability and environmental compatibility of the aquaculture (shrimp) industry.
- Zero Water Exchange Aquaculture Pond System (ZWEAPS) promises to advance aquaculture by providing better control over disease onset without harming the environment.

Biofloc Technology

Biofloc technology is a zero water exchange system where a high level of microbial bacterial floc is maintained in suspension using aeration. The addition of carbohydrates stimulates the growth of heterotrophic bacteria into the system, which uptake nitrogen directly from the ammonia released by the fish. Biofloc contains 25- 50% crude protein, 2.5% crude lipid, 4% fiber, 7% ash, and 22 KJ/g energy on a dry weight basis.

Bioflocs are conglomerates of microbes, macroaggregates composed of diatoms, macroalgae, protozoa, fecel pellets, exoskeleton, and remains of dead organisms, dead organic particles, bacteria, and invertebrates. Bioflocs found in ponds are porous, light, and have a diameter of 0.1 to a few mm. The co-culture of heterotrophic bacteria and algae is grown in flocs under controlled conditions within the culture tank or pond. Microbial biomass is grown on fish excreta, resulting in the removal of unwanted components from the water. The basic requirement of biofloc system operation is a high stocking density with 3 to 5 kg of fish or 130-150

PL10/m2, and high aeration either through a blower or with correct paddle wheel positions in ponds. Pellet grains and molasses are added to the culture water. Fish production can be enhanced up to 5 times than normal production, and shrimp production of 20-25 mt/ha/crop is normal for biofloc systems.

Biofloc in Feed and Feeding

Analysis of minerals indicates that microbial floc is rich in phosphorus and a wide range of other minerals. Biofloc contains different growth and probiotic factors, and it can replace the vitamins typically supplied in any intensive farming systems and in the feed for fishes. Feeding fishes like tilapia with bioflocs raises the specific activity of enzymes by about two- fold, possibly contributing to growth enhancement.

Bioflocs in Disease Management

The heterotrophic microbial biomass is suspected to have a controlling effect on pathogenic bacteria that are normally present in any aquaculture system. All intensive farming systems are heavily loaded with organic matter, principally of fish biomass, and therefore the chances for the outbreak of diseases are comparatively more. But the presence of biofloc can help in the control of a disease outbreak. Short-chain fatty acids act as biocontrol agents against pathogenic diseases. Poly-α-hydroxybutyrate (PHB), an intracellular biodegradable polymer produced by a wide variety of microorganisms, is involved in bacterial carbon and energy storage. This polymer depolymerizes in the fish gut and releases butyric acid, a known antimicrobial agent. As such, PHB might offer a prebiotic advantage for aquaculture. These findings suggest that biofloc can be a novel strategy for disease management on a long-term basis, in contrast to conventional approaches such as antibiotic, probiotic, and prebiotic application.

Periphyton Based Aquaculture System

Periphyton can be defined as a complex of sessile biota attached to submerged substrata such as stones and sticks, which includes algae, invertebrates, detritus, and microorganisms. Periphyton is considered an important food component for fishes, and nutrients are channeled through periphyton, which grows on various substrata in the aquatic environment and supports fish production. Depending on the type of substrata, periphyton communities are subdivided into 'Epilithon' (rock), 'Epipelon' (mud or silt), 'Epipsammon' (sand), and 'Epiphyton' (submerged macrophyte substrates). Any material providing surface area, including coral reefs, branches of different trees, higher aquatic plants, bamboo, PVC pipes, etc., can be used for periphyton production.

In the state of Manipur in the Northeastern part of India, substrate-based aquaculture systemsare widely prevalent in the Loktak Lake. Floating islands formed through the dense growth ofaquatic weeds and grasses are spread throughout the lake and are used as natural fish aggregating devices. These circular weed mats are locally called "Phums," and production obtained from these Phum areas is reported to be 300 to 1000 Kg/Phum.

Mechanism Involved in Periphyton based Aquaculture

- Two mechanisms have been proposed to account for higher levels of fish production in periphyton-based systems:
- The additional shelter provided by the substrata allows for a reduction in metabolic cost and a greater proportion of resources to flow into fish production.
- The additional primary and benthic production fostered by the artificial substrata support a new food web, part of which ends up in fish biomass.
- Additionally, filter feeding on phytoplankton is unlikely to provide the micronutrient and energy requirements of most herbivorous fishes, including rohu. Larger food items, such as benthic algae and algal detritus that grow on bamboo poles, are required to provide the additional complement of nutrients required by the fish.

Development of Periphyton Assemblage

The initial stages of periphyton development involve the deposition of dissolved organic substances on a clean surface, attracting bacteria through hydrophobic reactions. The presence of free-floating organic microparticles in eutrophic water further encourages this process. Bacteria actively attach using mucilaginous strands, with algae beginning to grow after a few days or a week. Low-profile diatoms are the first to appear, followed by small pennate diatoms, diatoms with rosettes, and mucilage pads, and finally, species of green algae with upright filaments or long strands.

Periphyton organisms utilize various mechanisms for attachment, including stalks with sticky ends, sticky capsules, cushions of filaments, muscular suction pads, glue, or simply clinging to the substrate. Attachment to sediment is achieved through rooting, rhizoids, or with a muscular foot.

As the density of organisms increases, competition for substrate surface area, carbon dioxide, nutrients, and light intensifies. This leads to changes in the composition of the periphyton community and explains its development away from the substrate.

Types of Substrate

Periphyton grown on substrates is known to improve water quality in aquacultureponds. Some of the common substrates used in Aquaculture are

- Bamboo pipe
- Plastic sheet
- Polyvinyl chloride(PVC) pipe

- Fibrous scrubber
- Ceramic tile

Factor Affecting Periphyton Growth

- Submersion time
- Substrate type
- Light intensity and quality
- Depth of the substrates

Periphyton Community

The most abundant family was Bacillariophyta (8 genera) followed by Chlorophyta (7 genera), and Cyanophyta (4 genera). Periphyton includes both the phyto periphyton and zoo- periphyton and sometimes aquatic insects. Common species constituting periphyton community is given below

Family	Species
Bacillariophyceae	Fragilaria, Diatom, Navicula, Cyclotella, Pinnularia
Myxophyceae/ Cyanophyceae	Oscillatoria, Rivularia, Nostoc, Anabaena
Chlorophyceae	Spirogyra, Scenedesmus
Eugleophyceae	Euglena
Rotifera	Branchionus angularis
Copepoda	Cyclops, Mesocyclops
Cladocera	Diaphnasoma

Role of Periphyton in Aquaculture

Food Source: Periphyton plays a crucial role in freshwater ecosystems by significantly contributing to primary productivity. It provides a vital energy source for both detritus and grazing food chains within the ecosystem. Periphyton is nutrient-rich and serves as a key food source for a variety of aquatic fauna, including fish, crustaceans, snails, chironomids, mayflies, and oligochaetes.

Efficient nutrient utilization: Periphyton can utilize nutrients already present in the water for their growth. Given their ability to grow on any substrate, they can be cultivated on readily available materials to support their growth.

Water Purification: Periphyton fulfills multiple functions in nutrient management by removing phosphate from the water column through uptake and deposition and by filtering particulate phosphate. These assemblages also contribute to nutrient retention by creating biochemical conditions conducive to phosphate deposition and by trapping particulate material from the water column..

Aquamimicry System

The concept of aquamimicry emerged in response to a shrimp disease outbreak in Thailand during the 1990s. It aims to recreate natural estuarine conditions by fostering zooplankton blooms, mainly comprised of copepods, as supplementary nutrition for cultured shrimp. Additionally, beneficial bacteria are used to maintain water quality. A carbon source, such as rice or wheat bran, is fermented by probiotics, such as Bacillus- like organisms, to release nutrients. This method is akin to the biofloc system, but with lower carbon input and less stringent nitrogen input regulations.

Preparation of Pond

The Aquamimicry system comprises three ponds: a sedimentation pond, a biofilter pond, and a grow-out pond. The grow-out pond should be filled with water to a depth of 0.80-1.00 m using a filter bag with a mesh size of 200-300μm. After adding probiotics (Bacillus) to the pond, dragging a chain along the bottom of the pond for a week enhances soil mixing with the probiotics and minimizes the development of biofilms that could be toxic to the shrimp. For lined ponds, heavy ropes should be used instead of chains to prevent tearing the liner. To eliminate small fish or eggs, piscicides can be applied at a concentration of ideally 20 ppm for teased cake or 50-100 ppm for fermented rice bran or wheat bran (without the husk). These additions help to promote copepod blooms, and additional rice bran can further flourish copepod bloom within two weeks. Aeration is recommended to ensure proper mixing of nutrients and probiotics in the pond and to reduce teased levels.

The sedimentation pond should be deep, reaching up to 4 m in the center and 2 m on the edges to allow for sediment accumulation. This pond can be stocked with bottom-dwelling fish species such as catfish or milkfish at low densities depending on the water salinity. The sediments from the grow-out pond encourage the production of worms and other benthic invertebrates that the fish can consume.

Water from the sedimentation pond overflows into the biofilter pond. This pond serves to increase water retention time while acting as a biofilter. Tilapia can be added at low densities in this pond. From here, the water overflows back to the grow-out pond with minimal nitrogenous waste. The sedimentation should be cleaned every three years. A major drawback of the system is that it requires a relatively large area, as the ratio of ponds is 1:1 (treatment to grow-out ponds).

Preparation of Carbon Source

For the proper functioning of this technology, a carbon source must be provided. Rice or wheat bran, without the husk, should be mixed with water at a ratio of 1:5-10 and probiotics under aeration for 24 hours. Finely powdered bran can be added slowly to the pond, whereas for the crumbled one, the upper "milk" or "juice" is added to the pond, and the bran solids are fed to fish in the biofilter pond. The pH of the incubation water should be between 6-7 and should be adjusted if necessary. Shrimp can be stocked at densities of 30-100 animals per square meter. As a general rule, the amount of fermented bran for extensive systems is 1 ppm, while for intensive systems it is 2-4 ppm, giving an ideal turbidity of 30-40 cm. If turbidity is higher, less bran should be added, and for lower turbidity, the amount of fermented bran can be increased. As the culture progresses, additional probiotics should be added each month to help maintain water quality and promote the formation of biocolloids (flocs composed of detritus, zooplankton, bacteria, etc.). To minimize the formation of biofilms, dragging with chains or ropes on the pond bottom is encouraged 15 days after pond stocking with shrimp. Further water quality management is not essential for extensive systems, but for intensive systems, excessive sediments should be removed to a sedimentation pond two hours after each feeding. It is reported that after harvesting, the pond bottom won't have a smell, black soil, or accumulated sediments.

Chapter-10

Running Water Culture System

Running water fish culture typically occurs in earthen ponds sized around 100 m^3 (10m x 10m x 1m) at high altitudes. Suitable fish species for this practice include trout, mahseer, and common carp. It's an intensive system requiring high stocking density and protein-rich feeds. Raceways, shallow tanks relying on high water flow, sustain aquatic life. Flow-through systems pass water through once, discharge it after waste treatment, and do not recirculate. Inflowing water temperature should match the species' tolerance. Oxygen is provided and removed by the fish as water progresses. Dissolved metabolites are carried out in the effluent, and waste is managed through settling or filtration. Nutritionally complete diets are essential. Flow-through systems, common in the U.S., are used for various coldwater species like trout and salmonids, as well as warm water species like catfish and tilapia. Raceways consist of tanks or series of rectangular tanks with water flowing along the long axis. Linear-pass raceways exhibit a water quality gradient from inflow to outflow, while circular units offer more uniform conditions. Raceways provide higher production per unit of space and better fish observation, feeding efficiency, disease detection, growth monitoring, and inventory estimates compared to ponds.

Different Types of Structures Used in Running Water Culture for Seed Rearing and Growing to Market Size

Raceways, with their diverse configurations such as semi-raceways, open or closed systems, long rectangular or circular channels, as well as flat spirals, are constructed using various materials such as wood, sheet metal, or concrete. Factors like topography, soil characteristics, erosion control, and water availability guide the selection of suitable sites for raceway installation. Construction involves tasks like leveling, excavation, and the installation of water control structures like weirs, aimed at regulating water depth and flow, and facilitating aeration. Water sources for raceways are carefully considered, with calculations made for flow rates, and closed systems often employ filtration and purification techniques for water quality maintenance.

Types of Raceway Ponds

- **Linear type:** Ponds are arranged sequentially, with a larger volume of water entering each pond. In this arrangement, disease occurrence in initial ponds may directly impact subsequent ponds due to the repeated use of the same water.
- **Lateral type:** Ponds are laid out in parallel, ensuring a fresh supply of water for each pond with a smaller volume entering each one. This layout prevents the transfer of disease from one pond to another while maintaining water quality.

Partitioned Aquaculture System (PAS)

In 1995, researchers at Clemson University constructed and operated prototypes of the Partitioned Aquaculture System (PAS) covering 0.13 hectares. This innovative technique enhances conventional fish pond function through the incorporation of slow-moving paddlewheel aerators and fish raceways. The primary advantage of the PAS, alongside high- rate pond systems, lies in its ability to boost algal production within the pond, facilitating ammonia assimilation and oxygen production.

The PAS concept involves partitioning fish pond culture into distinct physical, chemical, and biological processes interconnected by a uniform water velocity field. This physical separation serves two crucial purposes. Firstly, it enables precise quantification of fish culture and pond ecosystem processes. Secondly, by segregating fish culture operations from large-scale algal production and water treatment functions, it allows for independent optimization of each process, thereby enhancing overall performance.

The PAS design comprises three main physical components: algal channels with paddlewheels, fish raceways (with or without separate paddlewheels) separated from algal ponds by coarse screens, and a settling sump for capturing and concentrating settleable solids such as fish wastes and algal flocs for removal from the system.

The PAS capitalizes on the accelerated algal growth potential offered by high-rate pond systems. The system's driving force is the low-speed paddlewheel, which circulates water in a race track configuration. This innovative approach to mixed algal growth ponds had not previously been employed in aquaculture production. The velocity imparted by the paddlewheel ensures a well-mixed water column, enhancing exposure of algal populations to solar radiation and thereby increasing pond primary production. These features not only improve pond water quality but also provide biomass for algae byproducts or potential bioenergy resources.

Chapter-11

Recirculatory Aquaculture System (RAS)

A Recirculating Aquaculture System (RAS) is a method of aquaculture that involves treating and reusing water, with less than 10% of the total water volume replaced daily. Fish species such as channel catfish and rainbow trout have been effectively cultured using RAS.

One notable strategy for increasing production while minimizing waste is the implementation of recirculating aquaculture systems (RAS). These systems are engineered to collect and eliminate waste products, uneaten feed, and bacteria from the fish tank, allowing water to be recycled within the system. RASs can be configured for indoor or outdoor tanks or ponds, with variations in design and construction. However, most RASs perform key functions in wastewater treatment:

- Removal of solid waste.
- Removal or detoxification of nutrients.
- Removal of carbon dioxide.
- Supplementation of dissolved oxygen.
- Sterilization of bacteria and pathogens.

Solid waste particles, uneaten feed, and bacteria are concentrated and removed through settling or mechanical filtration, while fine particles (< 100 microns) are eliminated via foam fractionation or ozone treatment.

Dissolved nitrogenous wastes are addressed through biofilters, where wastewater is exposed to beneficial bacteria, facilitating nitrification and denitrification processes. Nitrification involves bacteria like Nitrosomonas sp. and Nitrosococcus sp. oxidizing ammonia into nitrite, followed by Nitrospira sp. oxidizing nitrite into nitrate. While ammonia and nitrite are toxic to fish in low quantities, nitrate can be tolerated at relatively higher levels. To mitigate potential harm from long-term nitrate exposure, many systems utilize a variety of microorganisms to facilitate denitrification.

Major Constraints for Operating RAS Systems in Aquaculture

Scaling up Recirculating Aquaculture Systems (RAS) to commercial operations poses significant challenges primarily due to the high costs linked with feed, labor, operations, and energy consumption. Moreover, the energy-intensive aspect of RAS, heavily reliant on electricity for powering recirculating systems, substantially increases its energy demands in comparison to other aquaculture systems. Additionally, issues arise regarding the accumulation of contaminants, fish mortality rates, and feed efficiency within RAS.

Species Suitable for RAS

- Barramundi/Asian Seabass/Bhetki (Lates calcarifer)
- Cobia (Rachycentron canadum)
- Silver/Indian Pompano (Trichinotus Blochii/Trichinotus mookalee)
- Tilapia (Oreochromis niloticus)
- Pearl spot/Karimeen (Etroplus suratensis)
- Pangasius (Pangasianodon hypophthalmus)

Components of RAS

- Insulated structure or building
- Storage and office space for feed and accessories
- Pump station
- Grow-out tanks: Circular cement or FRP tanks, complete with inlet, outlet, and central drainage
- Settling tanks for sludge management
- Water storage (sump) tanks
- Overhead reservoirs
- Mechanical (hydraulic) filters, drum filters, and glass wool/muslin cloth filters
- Pumping equipment and motors
- Power backup generator
- Sludge collection and solid waste management systems
- Biofilters and UV sterilization units
- Electrical infrastructure

- Automated feeding systems (as needed)
- Aeration and carbon dioxide removal systems
- Water quality testing equipment
- Water supply infrastructure, including bore wells
- Essential inputs such as seed, feed, additives, supplements, electricity/diesel, and labor.

Feed

- High-protein feed formulated with essential minerals and vitamins
- Tailored feed for specific fish species
- Feeding at a rate of 3-5% of the fish's body weight, adjusted according to the quality and protein content of the feed.
- Increased feeding frequency (multiple times daily) leads to enhanced growth rates and improved feed conversion efficiency.

Advantages of RAS

- Enhanced longevity of tanks and equipment.
- Decreased reliance on antibiotics and therapeutics, leading to higher-quality fish.
- Lower direct operational expenses related to feed, predator management and parasite control.
- Potential elimination of parasite release into surrounding waters.
- Reduced risks associated with climate, diseases, and parasite impacts.
- Flexibility in farming location and proximity to markets facilitated by RAS production.
- Ability to cultivate a diverse array of species regardless of temperature requirements.
- Improved feed management through continuous 24-hour monitoring in RAS.
- Minimization of stress exposure for fish due to adverse weather, temperature fluctuations, pollution, and predation.
- Secure production of non-native species.
- Efficient use of water and land resources.

Challenges of RAS

- Continuous and uninterrupted power supply is essential, necessitating backup electricity in case of power failures.
- The initial capital investment for establishing a recirculating aquaculture system is higher compared to ponds and raceways.

Chapter-12

Extensive and Semi Intensive Aquaculture System

This farming system is characterized by minimal management, utilizing large ponds spanning 1 to 5 hectares in size. Stocking density remains low, with fewer than 5000 fishes per hectare or more than 0.5 pieces per square meter for fish and more than 5 pieces per square meter for shrimp. There is no provision of supplemental feeding or fertilization. Harvest yields typically range from 500 to 2 tons per hectare, reflecting a lower productivity level. However, labor and investment costs are correspondingly low.

Extensive Aquaculture System in Brackish water

Traditional fish farming practices in India, such as bheri fish culture in West Bengal, fish and shrimp culture in pokkali paddy fields of Kerala, ghery fish farming in Chilka lagoon of Orissa, and Khazans of Goa, have been deeply ingrained for ages. These methods primarily involve polyculture, where fish seed from natural water bodies is introduced into ponds during high tide, without any additional input. Through auto stocking via sluice gates, fish seed entry occurs naturally during seasons of availability. Fish sustain themselves by feeding on natural diets within the ponds, eliminating the need for supplementary feed. Typically spanning 6-8 months, these practices yield between 500 to 1000 kg per hectare in a season, influenced by the diversity of fish species introduced and the abundance of natural diets within the pond.

Extensive Aquaculture System in Shrimp Culture

Extensive aquaculture systems adopt low stocking densities, typically ranging from 5,000 to 10,000 shrimp post larvae (PL) per hectare per crop, without the use of supplemental feeding. Fertilization may be employed to enhance the growth of natural food within the water. Water exchange occurs naturally through tidal mechanisms, with new water entering only during high tide and drainage taking place solely at low tide. These extensive culture ponds are often large, exceeding two hectares in size, and may retain natural features such as shallow depths and partial tree stump cover. Productivity remains modest, typically yielding less than 1 ton per hectare per year.

Semi Intensive Aquaculture System

This method utilizes relatively small ponds, ranging from 0.5 to 1 hectare in size, with higher stocking densities of 10,000 to 15,000 fish per hectare or 5 to 10 pieces per square meter for shrimp. Natural food is cultivated through fertilization, with optional supplemental feeding for fish, while shrimp receive supplemental feeding. Harvest yields typically range from 3 to 10 tons per hectare, resulting in a moderate level of productivity. Survival rates are notably high in this approach.

Semi Intensive Aquaculture System in Brackishwater

This method involves careful selection of hatchery-reared seed and the application of formulated pellet feed. Typically, fish stocking occurs at a rate of 6000-8000 individuals per hectare. Following pond preparation, feeding commences with artificial diets, alongside diligent management of water quality. Adjustments in water exchange are made as fish grow and biomass increases. Installation of aerators is essential to enhance dissolved oxygen levels in the pond. All activities are meticulously scheduled to maximize production. Through improved pond culture techniques, fish production levels of 4 to 6 tons per hectare can be achieved.

Semi Intensive Aquaculture System in Shrimp Culture

Semi-intensive aquaculture systems employ higher stocking densities compared to extensive systems, ranging from 50,000 to 100,000 shrimp post larvae (PL) per hectare per crop, accompanied by supplementary feeding. In contrast, semi-intensive culture relies entirely on artificial, formulated feeds, with densities reaching as high as 200,000 to 300,000 shrimp PL per hectare per crop. Both systems utilize small pond compartments up to 1 hectare in size for easier management.

Management of semi-intensive systems involves input application, including feeds, fertilizers, lime, and pesticides, alongside environmental manipulation primarily through water management using pumps and aerators. Feeding occurs at regular intervals throughout the day, with computed daily feed rations administered in equal doses, typically ranging from three to six times daily. Daily water exchange replenishes approximately 10-15% of the pond volume in semi-intensive shrimp culture.

Semi-intensive culture systems demand more labor compared to extensive systems and entail higher setup and operational costs, as well as increased risks of mortality due to disease and poor management. However, production levels are significantly higher, with semi-intensive shrimp culture yielding at least 1.5 tons per hectare per crop. Although financial returns are more attractive compared to extensive culture, the return on investment (ROI) from semi- intensive culture surpasses that of intensive culture due to the elevated costs associated with intensive systems, primarily fry and feeds.

Extensive/ Semi Intensive Aquaculture Systems in India

Pen Culture

Pen culture involves confining fish in a designated water volume, enclosed on all sides except the bottom, enabling water circulation from at least one side. Combining aspects of pond and cage culture, this method typically employs net or wooden materials to create pens along shallow regions of lakes and reservoirs' shores and banks. These pens utilize the lake's bottom, which harbors benthic fauna as fish food. Polyculture, akin to pond culture, can be practiced in these pens, which maintain a water environment with ample dissolved oxygen due to free water exchange with the surrounding body. This method finds extensive use in Japan, Peru, and the Philippines, where, for example, milkfish fingerlings are reared in pens in Laguna de Bay until reaching marketable size (200 g or more). Typically constructed using split bamboo or netting, these pens house fish from post-monsoon to late autumn, with a culture period spanning 6 to 8 months. Notably, major carp seed rearing has been an annual practice since 1982 in the Tungabhadra reservoir in Karnataka.

Pond Culture

This method of fish culture involves maintaining water in enclosed areas through artificially constructed ponds, where various aquatic animals like finfish and shellfish are raised. Ponds can be filled with canal water, rainwater, borewell water, or other water sources, with proper consideration given to site selection based on climate, topography, water availability, and soil quality.

There are two primary types of ponds based on these factors:

1. Watershed Pond Systems

- These ponds rely entirely on watershed runoff to fill and maintain water levels, although groundwater and surface water can supplement the supply.
- Factors such as soil type, topography, and water quality are crucial considerations.

- Soil type, particularly mud or mud-silt soils, is important for preventing leakage.
- Topography dictates pond size and shape, with a maximum depth of 10 feet recommended.
- Management is more challenging due to erratic water supplies and uneven pond bottoms.

2. Levee Pond Systems

- Constructed in flat land areas where watershed runoff is insufficient, these ponds are typically filled with groundwater.
- Soil characteristics, groundwater availability, and topography are essential factors for site suitability.
- Adequate clay content (at least 20%) in the soil prevents leakage.
- Rectangular pond shapes are preferred for simplicity and ease of management.
- Water quality considerations are vital for successful pond construction.
- Levee ponds are more suitable for fish growth and harvesting compared to watershed ponds.

3. Hybrid Pond

These ponds represent a hybrid approach, blending features of both embankment and watershed ponds. They typically feature embankments on two or three sides, enclosing a relatively small, shallow drainage basin. Hybrid watershed-embankment ponds are constructed in areas with gently rolling topography, where neither pure embankment nor watershed ponds are ideally suited.

Tank Culture

Utilizing tanks instead of open ponds, tank culture is a controlled method of raising aquatic animals such as fish and shellfish. It presents numerous advantages over traditional pond culture, rendering it a favored option for various species and applications.

Advantages of Tank Culture

Tank culture offers enhanced environmental control, with precise management of water quality parameters such as temperature, pH, and dissolved oxygen, along with control over lighting and feeding. This meticulous oversight leads to optimized growth and survival rates. Additionally, tanks allow for higher stocking densities, maximizing production per unit area compared to ponds. Moreover, they require significantly less water, promoting sustainability, particularly in water-scarce regions. The closed systems in tanks also improve biosecurity by minimizing the risk of disease outbreaks and the introduction of harmful pathogens. Furthermore, controlled conditions can accelerate growth rates, resulting in faster harvests and shorter production cycles. Unlike traditional methods reliant on seasonal weather fluctuations, tank culture enables year-round production. Additionally, the ease of monitoring and management afforded by tanks facilitates closer observation of fish health and behavior, enabling timely interventions when necessary.

Chapter-13

Intensive and Super Intensive Aquaculture System

Intensive fish farming represents a meticulously managed approach aimed at maximizing fish production within limited water resources. This system employs small ponds, tanks, or raceways with exceptionally high stocking densities ranging from 10 to 50 fish per cubic meter of water. Fish are exclusively fed formulated feed, and meticulous management practices are implemented, including proper aeration using aerators and the provision of highly nutritious feed.

Yields obtained from intensive fish farming typically range from 15 to 100 tons per hectare or even more, showcasing significantly higher productivity compared to other farming systems. However, the initial investment costs are notably high due to the advanced infrastructure and technology required. Despite this, the substantial yields and efficient resource utilization make intensive fish farming a highly profitable venture.

Intensive Aquaculture in Brackishwater

- This approach involves the careful selection of hatchery-reared seed and the use of formulated pellet feed. Typically, fish stocking occurs at a rate of 6000-8000 individuals per hectare. Following pond preparation, feeding commences with artificial diets, alongside diligent management of water quality. Adjustments in water exchange are made as fish grow and biomass increases, while the inclusion of aerators is essential to enhance dissolved oxygen levels in the pond.
- All tasks are meticulously scheduled to optimize production levels. Through the improved pond culture method, fish production levels ranging from 4 to 6 tons per hectare can be achieved.

Intensive Aquaculture in Shrimp Culture

Intensive culture employs exceptionally high densities of cultured organisms, such as 200,000 to 300,000 shrimp post larvae (PL) per hectare per crop, relying solely on artificial, formulated feeds. Both systems utilize small pond compartments, typically up to 1 hectare in size, for simplified management.

Management of intensive culture systems involves the application of inputs, primarily feeds, fertilizers, lime, and pesticides, alongside manipulation of the environment primarily through water management using pumps and aerators. Feeding occurs at regular intervals throughout the day.

In intensive shrimp culture, the daily feed ration is administered in equal doses, ranging from three to six times a day. Daily water changes, with approximately 10-15% replenishment, are essential for maintaining water quality.

Intensive culture systems demand more labor compared to extensive systems, are costlier to establish and operate, and carry higher risks of mortalities due to disease and poor management. However, production levels are significantly higher, with intensive shrimp culture yielding a minimum of 10 tons per hectare per crop.

While financial returns are more attractive compared to extensive culture, studies indicate that the return on investment (ROI) from semi-intensive culture may be better than that of intensive culture due to the high costs associated with inputs, primarily fry and feeds, in intensive systems.

Super Intensive Aquaculture System

- Super-intensive aquaculture relies on continuous water flow, with daily water exchange being standard practice.
- This system is predominantly implemented in cement tanks, fiberglass tanks, and raceways, all equipped with high-efficiency biological filters for continuous water recirculation. The tank sizes typically range between 50-100 m^3.
- High-quality formulated feeds are provided to the cultured organisms via demand feeders. Water quality is meticulously monitored using electronic devices.
- Stocking density ranges from 800,000 to 1,000,000 shrimp PL/ha or 40,000 to 50,000 carp fingerlings/ha.
- Production levels typically range from 10 to 12 tons/ha in shrimp culture and 15 to 20 tons/ha in carp culture.

Intensive Aquaculture System in India

Cage Culture

Cage culture, a method of fish farming, involves raising fish from juvenile to commercial size in a confined water volume, enclosed on all sides, including the bottom, while allowing water circulation. It's particularly suited for areas like lakes, reservoirs, and riverine pools where draining isn't possible. Widely practiced in countries such as Japan, South Korea, China, the Philippines, Thailand, Malaysia, Germany, Norway, and the USA, virtually all cultivable fish species, like carps, tilapia, trout, and catfish, can be successfully raised in cages, depending on socioeconomic, ecological, and technical factors. Cost-effective and simplifying fish health monitoring, cage culture enables complete harvesting without fish escaping. It's typically practiced in water bodies with minimum depths of 5 meters in reservoirs, 3 meters in wetlands, and 2 meters in canals. Commonly cultured freshwater fish and shellfish suitable for cage culture include Indian Major Carps, Chinese Carps, air-breathing catfishes, climbing perch, tilapia, snakeheads, and freshwater prawns.

Biofloc Technology

Biofloc technology operates as a water-efficient system with minimal exchange, relying on aeration to sustain a dense microbial bacterial floc. Introducing carbohydrates encourages the proliferation of heterotrophic bacteria, which efficiently utilize the ammonia produced by fish, directly contributing to nitrogen uptake. On a dry weight basis, biofloc typically comprises approximately 25-50% crude protein, 2.5% crude lipid, 4% fiber, 7% ash, and offers an energy content of 22 KJ/g.

RAS (Recirculatory Aquaculture System)

A Recirculating Aquaculture System (RAS) is a method of aquaculture that involves treating and reusing water, with less than 10% of the total water volume replaced daily. Fish species such as channel catfish and rainbow trout have been effectively cultured using RAS.

One notable strategy for increasing production while minimizing waste is the implementation of recirculating aquaculture systems (RAS). These systems are engineered to collect and eliminate waste products, uneaten feed, and bacteria from the fish tank, allowing water to be recycled within the system. RASs can be configured for indoor or outdoor tanks or ponds, with variations in design and construction. However, most RASs perform key functions in wastewater treatment:

- ❑ Removal of solid waste.
- ❑ Removal or detoxification of nutrients.
- ❑ Removal of carbon dioxide.
- ❑ Supplementation of dissolved oxygen.
- ❑ Sterilization of bacteria and pathogens.

Solid waste particles, uneaten feed, and bacteria are concentrated and removed through settling or mechanical filtration, while fine particles (< 100 microns) are eliminated via foam fractionation or ozone treatment.

Dissolved nitrogenous wastes are addressed through biofilters, where wastewater is exposed to beneficial bacteria, facilitating nitrification and denitrification processes. Nitrification involves bacteria like Nitrosomonas *sp.* and Nitrosococcus *sp.* oxidizing ammonia into nitrite, followed by Nitrospira *sp.* oxidizing nitrite into nitrate. While ammonia and nitrite are toxic to fish in low quantities, nitrate can be tolerated at relatively higher levels. To mitigate potential harm from long-term nitrate exposure, many systems utilize a variety of microorganisms to facilitate denitrification.

Aquaponics System

In the domain of aquaponics, effluents from Recirculating Aquaculture Systems (RAS) serve as a nutrient-rich resource for plants (Paudel, 2020; Wongkiew *et al.*, 2017). However, the implementation of these systems can be expensive due to the installation requirements for both RAS and hydroponic systems (Palm *et al.,* 2018). Plants that thrive in hydroponic environments typically excel in aquaponic setups. If the nutrient levels in the RAS effluent are inadequate, additional fertilizers can be applied (Eck *et al.*, 2019; Maucieri *et al.*, 2019).

Decoupled aquaponic systems allow for the adjustment of water flow through fish tanks, filters, and hydroponics, unlike coupled systems where the water flows through all components (Goddek *et al.*, 2016; Monsees *et al.*, 2017). The potential of aquaponic systems lies in their capacity to minimize nutrient losses from aquaculture. A recent innovation is the flocponic system, which replaces RAS with a biofloc system linked to hydroponics (Pinho *et al.*, 2017).

Key challenges for profitability in aquaponic or flocponic systems include the need to access high-value markets for vegetables and fish, the diverse knowledge base required, and the significance of vegetables in ensuring financial success (Bosma *et al.*, 2017). These factors contribute to the slow adoption of these systems in the industry.

Partitioned Aquaculture System (PAS)

In 1995, researchers at Clemson University constructed and operated prototypes of the Partitioned Aquaculture System (PAS) covering 0.13 hectares. This innovative technique enhances conventional fish pond function through the incorporation of slow-moving paddlewheel aerators and fish raceways. The primary advantage of the PAS, alongside high- rate pond systems, lies in its ability to boost algal production within the pond, facilitating ammonia assimilation and oxygen production.

The PAS concept involves partitioning fish pond culture into distinct physical, chemical, and biological processes interconnected by a uniform water velocity field. This physical separation serves two crucial purposes. Firstly, it enables precise quantification of fish culture and pond ecosystem processes. Secondly, by segregating fish culture operations from large-scale algal production and water treatment functions, it allows for independent optimization of each process, thereby enhancing overall performance.

The PAS design comprises three main physical components: algal channels with paddlewheels, fish raceways (with or without separate paddlewheels) separated from algal ponds by coarse screens, and a settling sump for capturing and concentrating settleable solids such as fish wastes and algal flocs for removal from the system.

The PAS capitalizes on the accelerated algal growth potential offered by high-rate pond systems. The system's driving force is the low-speed paddlewheel, which circulates water in a race track configuration. This innovative approach to mixed algal growth ponds had not previously been employed in aquaculture production. The velocity imparted by the paddlewheel ensures a well-mixed water column, enhancing exposure of algal populations to solar radiation and thereby increasing pond primary production. These features not only improve pond water quality but also provide biomass for algae byproducts or potential bioenergy resources.

Flocponics System

Flocponics represents an innovative aquaculture approach that merges the principles of recirculating aquaculture systems (RAS) with hydroponics. It entails cultivating fish within a controlled environment where water is continuously recirculated to maintain optimal conditions for fish growth. The waste generated by the fish is transformed into ammonia, which is further metabolized by nitrifying bacteria into nitrate. This nitrate-rich water is then conveyed to a hydroponic setup, where plants are grown without soil. These plants utilize the nitrate as a nutrient source, effectively purifying the water. This creates a mutualistic relationship between the fish and plants, with the fish providing nutrients for the plants, while the plants aid in filtering the water for the fish. Flocponics offers numerous advantages over traditional aquaponics systems, including enhanced nitrate removal, improved water quality, minimized environmental footprint, increased adaptability, better fish health, and reduced labor and energy expenditures.

Chapter-14

Site Selection for Construction of Aqua Farm

The selection of an appropriate site for fish farming is crucial to its success. Two main factors that determine the suitability of a site for pond construction are the soil's water retention capacity and its inherent fertility, as well as its ability to respond well to both organic and inorganic fertilization. Additionally, a dependable, perennial source of water is essential to fill the ponds throughout the year. Ideally, the site should have a gentle slope to allow for the construction of self-draining ponds at higher elevations. In swampy or marshy areas, bunds must be built up with deposited earth to create ponds of the required size. Accessibility by road or rail and proximity to markets for easy fish disposal are also important. Nearby availability of necessary inputs like feeds, fertilizers, construction materials, and other accessories is crucial. The site should be free from industrial, domestic, insecticide, and pesticide pollution. Furthermore, the farm should offer opportunities for integrating aquaculture with agriculture and animal husbandry. Site selection also depends on the type of fish intended for farming.

When selecting a site and constructing ponds, these factors should be taken into consideration.

1. Ecological Factors

- Soil
- Water

- Land
- Climate

2. Biological and Operational Factors

- Identification of the species to be cultivated and assessment of the resources and availability of stocking materials such as spawners, fry, or fingerlings.
- Determination of the project type, whether it is a small-scale rural project or a large-scale rural project.
- Selection of the culture system adopted, which could be extensive, semi-intensive, or intensive.
- Specification of the operational method, including monoculture, polyculture, or integrated approaches.
- Establishment of production targets for the project.
- Estimation of the required area size for project implementation.

3. Economic and Social Factors

- Evaluation of land ownership, land availability, land values, regulations, and legal constraints regarding land use within the project area.
- Assessment of proximity to all-weather road networks to ensure efficient transportation.
- Examination of access to electricity, telephone, radio connections, and associated utility costs.
- Analysis of equipment, services, and supplies availability necessary for project operations.
- Evaluation of the availability of construction materials required for infrastructure development.
- Identification of market locations and demand determination for project produce.
- Assessment of the availability of organic and synthetic fertilizers, medications, and chemical inputs.
- Examination of the availability of supplementary feeds required for project operations.
- Analysis of the costs associated with equipment, materials, and feeds essential for project implementation.
- Review of suitable transportation facilities for product distribution.
- Assessment of ice availability for marketing purposes.
- Evaluation of staff availability with adequate experience in pond management techniques.

- Identification of skilled and semi-skilled labor availability within the project area.
- Consideration of amenities for permanent staff members, such as schools, shopping facilities, and healthcare services.
- Investigation into local financing methods or credit opportunities.
- Examination of political dynamics and realities within the project area.

Ecological Factors

Soil

The quality of the soil plays a critical role in both the productivity and water quality of a pond, as well as its suitability for dike construction. When assessing soil suitability for pond construction, the most important factors to examine are soil texture (particle size composition) and porosity or permeability (the ability to allow water to pass through). The pond bottom must be capable of holding water, and the soil should also contribute to water fertility by providing nutrients (Clay soil, for example, is typically more fertile). Therefore, the most ideal soil for pond construction should contain a high percentage of clay. Generally, silt clay, clay-loam, and loamy soils are considered suitable for fish ponds. A good quality soil should consist of 90% soil fraction and not exceed 10% in gravel content. It is advisable to avoid rocky, sandy, gravel, and limestone areas. If dealing with porous soil, the pond bottom can be treated with bentonite, clay, or other soil sealants. Alternatively, plastic film liners can be used to reduce the rate of seepage. However, such treatments, while effective, can be expensive and may hinder the exchange of soil-water minerals and nutrients, thereby making the pond less productive. To assess whether the soil is suitable for pond construction, one should consider these three key factors.

- Squeeze method
- Ground water test
- Water permeability test

a. Squeeze Method

To determine the suitability of the soil for pond construction, take a wet handful of soil and firmly squeeze it. If the soil retains its shape after you open your hand, it is considered suitable for pond construction.

b. Ground Water Test

To ensure accurate results, conduct this test during the dry season. Dig a one- meter- deep hole and cover it with leaves overnight. If the hole is filled with groundwater the next morning, the site may be suitable for pond construction, although draining may take longer due to high groundwater levels. If the hole remains empty the next morning, then the site is likely suitable for farming. The next step is to test the soil for its water permeability.

c. Water Permeability Test

To conduct the permeability test, fill the hole with water up to the top and cover it with leaves. The next day, if the water level decreases, it indicates more water seepage in the soil. To reconfirm this, refill the hole with water to the top and cover it with leaves again. Check the water level the following day. If the water level remains high, the soil is impermeable enough and is suitable for pond construction.

However, if the water has disappeared again, the site is not suitable for fish farming unless the bottom is first covered with plastic or heavy clays.

Water

Availability of an adequate and dependable source of water is another prerequisite for setting up a fish farm. The pond should be filled with water at regular intervals so as to adjust water depth. The usual source of water supply to fish farm are reservoirs, streams, springs, canals, surface run off (rain), wells, tube wells etc. However, the best thing to do is to have as far as possible natural, preferably rain Water and other large water bodies, because in natural 'water bodies, dissolved oxygen content, pH, water quality and water temperature are more stable and suitable for fish to grow. On the contrary, the water from underground source is often deficient in dissolved oxygen. The waste water discharged from factories and mines usually contain harmful chemicals which are not suitable for fish farming. The ponds should be at a safe distance away from rivers and reservoirs so as to avoid water flooding during rainy season. Water should not be either highly acidic or alkaline. It should be corrected if acidic by the application of lime and if alkaline by adding organic manure. Water parameters should also be taken into consideration in selection of an area for faming

- **Water temperature:** The water temperature is an important condition in assessing whether the fish species selected can be raised. A water temperature between 20°C and 30°C is generally good for fish farming.
- **Water salinity:** Variation in water salinity (amount of dissolved salts in the water) is also an important factor which must be considered. Some fish species can withstand a wider salinity range than others: e.g. tilapia and catfish can withstand a wide range from fresh- to seawater while carp can only withstand freshwater.

Land (Topography)

Topography of the land determines the type of pond to be constructed. The land selected for farming should be with slope not steeper than two percent. The land should be free from high flooding; the high flooding recorded for the past ten years should not be higher than the top of the dyke planned for the pond construction. he land with regular shape is more suitable for farm construction. A with scope for industrialization which may cause pollution in future should be avoided. Likewise area already located with industries must be avoided. Areas having underground oil pipe lines, high electricity power poles and radio masts are not considered suitable for farm. Lands with heavily rooted vegetation are also not suitable for construction of farm.

Climate

The following important climatological factors are to be collected from the metereological station or from the past incidences from the selected area for farm construction

- Temperature
- Rainfall
- Evaporation
- Humidity
- Sunshine
- Wind speed and direction
- Incidence of high rainfall, wind/heavy storms and earthquake

Biological and Operational Factors

Before initiation or finalization of a site for farm construction the following factors are to be ascertained

- Species to be cultured
- Availability of seed for farming
- Type of project based on investment
- System of culture to be adopted

Based on these biological factors only an area has to be selected for farming

Economic and Social Factors

Land without any legal problems should be selected; the selected area for farming should not have any local problems or conflicts with the local population residing. The selected site should have road facility which is accessible in all weather conditions, availability of electricity, communication facility (Telephone or mobile), availability of construction material, equipments, labour (skilled and semi skilled), staff, hospital, school, bank, transport etc and all other facilities required to run a farm.

Construction

Before construction of a pond a proper design and lay out should be prepared. The size and shape of the pond should be designed in such a way that the excavated earth should be used for construction of dyke. The land is to be excavated with a gradual slope towards the outlet for proper draining facility. Construction of pond must be planned in such a way that, construction must get completed in summer so the pond can be used for stocking.

Different Kinds of Ponds and its Requirement Ratio in a Farm Unit

Fish culture commences with the construction of ponds. Specific types of ponds are required based on species of fish and various life history stages. Various ponds like nursery, rearing, stocking and treatment ponds are required for a fish farm. The ratio of these ponds varies in relation to farming practice, stocking density and survival.

The ideal shape of the pond should be rectangular. Length and breadth in the proportion of 3:1 is ideal. In any case, the breadth should not be more than 30-50 meter as it is difficult to operate efficiently a net longer than 50 meter. The corners of the pond should be rounded, so that while operating net, fish may not escape. In Indian conditions the total fish farm should be constructed consisting of nursery pond (5% of the total area), rearing pond (20% of the total area), stocking ponds (70% of the total area) and bio pond or treatment pond (5% of the total area).

Size and Depth of the Different Kinds of Pond

Other factor involved for site selection, is the purpose of farm. The purpose of the farm depends how large is the size of the area. A Scientifically constructed fish farm has 3 types of ponds, which comprise of nursery, rearing, stocking, brooder pond and breeding ponds etc., a large area is required.

Nursery Pond

The nursery ponds, each with an ideal size ranging from 0.01 to 0.05 hectares, are shaped in a regular rectangle with a flat, even bottom to facilitate the ease of netting operations. The ponds' water depth is maintained at 1-1.5 meters, and the number of ponds varies according to the targeted production level.

Rearing Pond

In a rearing pond, fry are nurtured to the fingerling stage, measuring 10-15 cm in length, over a period ranging from 2-3 months. Each fingerling pond, which ranges in size from 0.05 to 0.1 hectares, has a water depth of 1.5-2 meters and is designed with a rectangular shape and a flat, even bottom to facilitate straightforward netting operations. The number of fingerling ponds is adjusted according to the desired production of fingerlings.

Stocking Pond

The stocking ponds are of an ideal size, ranging from 1 to 2 hectares, for the intensive cultivation of food fish. The water depth is maintained at 2.5 to 3 meters, allowing the ponds to accommodate fingerlings measuring 10-15 cm in length, which are nurtured until they reach marketable size. The number of stocking ponds is adjusted according to the target fish production, and the rearing period typically ranges from 8-10 months. The ponds may also serve as breeding or brooder raising ponds, although there is no rigid requirement for their size.

Bio Pond or Treatment Ponds

A notable addition to modern fish farming practices is the inclusion of a specific pond type known as a Bio pond. These ponds serve as large settling tanks, playing a vital role in the biological purification of water used in the farm's fish ponds. When required, Bio ponds can also double as stocking ponds. Typically, the area occupied by a Bio pond constitutes 7-10% of the total productive area of the fish farm.

Positioning of Different Kinds of Ponds in a Fish Farm

Strategic pond placement within a fish farm is crucial for streamlined operation and cost efficiency. Conducting a thorough contour survey of the site before construction helps minimize the need for filling, digging, or earth movement, reducing labor and mechanical costs.

The lower, deeper sections of the site should be designated as stocking ponds, while higher areas are ideal for rearing ponds. The highest points of the property are best suited for office buildings, staff quarters, laboratories, and hatcheries.

Types of Ponds in Fish Farm

Dug-out ponds, excavated in flat terrain, offer fish farmers the flexibility to tailor ponds to their specific requirements, ensuring optimal shape, size, and depth based on scientific standards.

Embankment ponds, commonly found in hilly regions, involve constructing dykes on one or two sides of the chosen area. While digging out ponds directly from the site may seem cost-effective, it often fails to meet ideal fish culture standards due to limited control over size, shape, and depth. In embankment ponds, dykes are raised to suitable heights for fish culture, with inlets and outlets provided. To regulate entry and prevent escape of cultured fish, small mesh bamboo or nylon screens are installed at these openings, effectively deterring unwanted aquatic organisms from entering the culture system.

Steps to be Followed in Construction of a Fish Pond are

- Begin by clearing the site, removing trees, bushes, and rocks, and marking the area designated for pond construction.
- Next, construct a clay core to establish a leak-proof and durable dyke.
- Afterward, excavate the pond and build the dyke on top of the clay core.
- Install the inlet and outlet structures to regulate water flow.
- Protect the dyke by covering it with soil and planting grass such as Rhodes grass (*Chloris gavana*) or Star grass (*Cynodon dactylon*). Avoid using long-rooted trees or plants, as they can weaken the dyke.
- Finally, fence the pond to deter theft and protect against predatory animals.

Dyke

To ensure the effective management of water and its fertility, the dyke around a fish pond must be solid, compacted, and leak-proof, offering stability under all weather conditions. It is vital that the dyke's height exceeds the maximum flood water level by 0.3 to 0.7 meters. For constructing the dyke, a soil composition containing 15-30% silt, 45-55% sand, and 30-35% clay is optimal.

A berm, providing ample width to stabilize the slope and facilitate net operation in the pond, is essential. The berm's width must not be less than 1 meter and should be ideally wider. A 2:1 slope is recommended for embankments in good quality clay soil, while a 3:1 slope is suitable for loamy silt or sandy soils.

When raising the dyke, a clay puddle, comprising a mixture of 1 part sand and 2 parts clay, should be deposited in 10-15 cm thick layers, ensuring adequate moistening between layers. The puddle may be placed at the center or on the water side of the pond. The crest of the dyke should be broad enough to accommodate allied farm activities gradually, with a minimum top width of 1 meter.

To mitigate soil erosion, short creeping grass should be grown on the top and sides of the dyke. On broader embankments, banana and coconut trees can be planted to avoid shading during the morning hours.

Additionally, terrestrial grasses like hybrid Napier, gunny grasses, and elephant grasses can be cultivated on wider crests and slopes to provide feed, either wholly or partially, to herbivorous fishes like grass carp reared in cultured tanks.

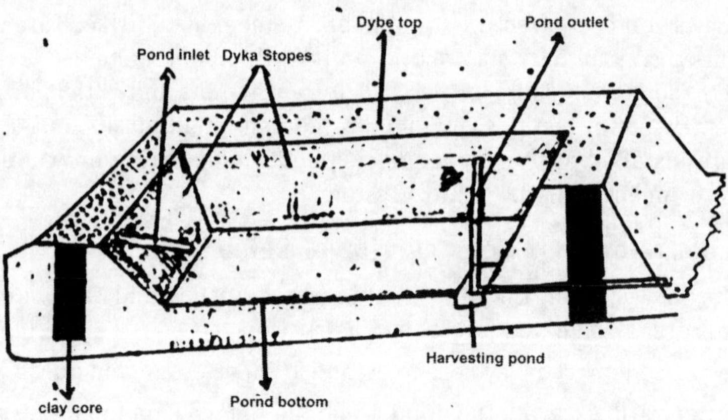

Fig.1: Dyke Construction in Fish Pond

Inlet canals

Feeder canals play a crucial role, particularly for ponds not directly filled by spring or rainwater. These canals must be designed to deliver the necessary quantity and quality of water to the pond. To ensure optimal functionality, the inlet pipe should be positioned above the maximum water holding level of the pond and equipped with a screen to filter the pumped water.

Standard inlet pipe size used based on the size of the pond

Pond Size (M^2)	Inside Diameter of Pipe (Cm)
<200	not less than 10
200-400	10-15
400-1 000	15-20
1 000-2 000	20-25
2 000-5 000	25-30
>5 000	40 or more

Outlet Canals

The outlet system, or drainage canal, serves the purpose of draining the pond during harvest or for partial water exchange to maintain water quality during the culture period. It is essential to construct the outlet before building the pond dyke.

Different Outlet Canals or Draining Systems

a. Rivaldi Valve

Named after the innovative farmer from Paraguay who pioneered its use, the Rivaldi valve is a flexible pipe system employed for drainage purposes. Before constructing the dyke, the flexible pipe is positioned on the ground, with one end turned up and secured to a stake or pole above the pond's maximum filling level. During draining operations, the stake is removed, and the pipe is laid down, facilitating the outflow of excess water, particularly during periods of heavy rainfall.

b. Sluice Gate

A sluice can serve dual functions as both an inlet and outlet canal, depending on its placement within the pond. Typically utilized as an outlet system, a sluice is a cement structure comprising various slot or groove arrangements sealed by wooden gates. These wooden gates, often constructed from multiple pieces, are inserted into the grooves and filled with earth to secure them in place. Additionally, a wooden-framed meshed gate is installed in one of the grooves to prevent fish from escaping during water exchanges.

c. Monk

Similar to a sluice but situated within the pond dyke, a monk serves a distinct purpose and cannot function as an inlet like a sluice. It comprises a horizontal drainage pipe and a vertical structure, enclosed on three sides and open on one. The monk structure features grooves along its two parallel sides and bottom, which accommodate wooden and meshed gates to regulate water flow.

Maintenance

During fish culture operations, sediment and organic debris tend to accumulate on the pond bottom, necessitating periodic removal every 3 to 4 years. In instances of excessive silt buildup, desilting is recommended, followed by exposing the dry pond bottom to sunlight through drying. Annual maintenance of the pond dyke post-monsoon is essential, involving leveling the dyke crest, thorough rolling, and planting grass to enhance stability. Any holes formed in the pond sides or embankments should be filled with clay or loam to prevent erosion.

Chapter-15

Carrying Capacity of Fish Pond

Carrying capacity, alternatively termed stocking density or stocking rate, denotes the quantity of fry or fingerlings per unit area of water. Typically expressed as the number or weight of fish per square meter (m^2), it represents the number of fish to be stocked relative to the size of the pond.

It is advisable to stock fish following the pond's carrying capacity aligned with the desired market size. This is due to the pond functioning as an ecosystem with a finite capacity to sustainably support aquatic life. When stocked beyond its capacity, the pond may transition from a conducive production environment to a polluted system incapable of supporting optimal production. As a consequence of overloading, the pond's water quality begins to decline as its natural systems struggle to effectively process and manage waste.

Upon reaching the pond's carrying capacity, fish growth ceases irrespective of the duration spent in the pond, leading to diminishing production and returns, commonly referred to as the "point of diminishing returns."

The Pond's Carrying Capacity is Affected by Various Factors

- Fish size within the pond, as it directly impacts the rate of feeding.
- The species of fish being cultivated; for instance, air-breathing fish have a higher carrying capacity compared to carps, as they do not solely depend on dissolved oxygen in the pond.
- The quantity and composition of feed or fertilizer introduced into the pond.
- The volume and quality of water present in the pond.

Stocking in Nursery Pond

Spawn measuring approximately 47 mm in length are typically introduced into the nursery pond, where stocking rates are usually high. The recommended optimal stocking density ranges from 6 to 10 million spawn per hectare.

Stocking in Rearing Pond

Typically, advanced fry measuring between 20-30 mm are stocked in rearing ponds at a density of 1.5 lakh per hectare.

Estimation of Stocking Density

Estimating the stocking density for a pond can be determined by assessing its carrying capacity.

$$\text{Number to stock} = \frac{\text{Estimated maximum carrying capacity of a pond(Kg)}}{\text{Desired weight at harvest (Kg)}} + 10\%$$

Chapter-16

Pre-Stocking Management in Carp Farming

Pre-stocking management encompasses all the preparatory measures undertaken in fish culture prior to the introduction of fry into the water body, aimed at optimizing the aquatic environment and surrounding conditions for the healthy growth and development of the fry.

Carp Seed Raising

The initial phase of seed raising involves nurturing spawn from three to four days old to the fry stage (5-6 mm in length) of cultivable carps in a designated nursery pond.

During this stage, the hatchlings have completed the yolk absorption process and begin to feed autonomously from external sources.

The spawns are then transferred and stocked into a nursery pond that has been prepared in advance, typically characterized by smaller dimensions (0.01-0.05 ha) and water depths ranging from 1 to 1.5 meters.

Nursery Pond Management

Steps Involved in Pre Stocking Management in NurseryPond

- ❑ Preparing the pond
- ❑ Eliminating aquatic weeds and insects from the fish pond
- ❑ Applying lime to the pond

- Filling the pond with water
- Providing basal manure and fertilization

Pond Preparation

Pond preparation is the main function of Pre-stocking management.

Pond Drying / Dewatering

Drying the pond is a crucial aspect of physical development during pond preparation, serving several key objectives:

- Eliminating undesirable species and predatory fishes from the pond.
- Facilitating dike and predator fish management efforts.
- Aiding in the application of lime and fertilizers to the soil.

It is recommended to allow the pond bottom to dry under sunlight for a minimum of 15 days.

Desilting /Bottom Mud Excavation

Following dewatering of the pond, it is imperative to remove excess bottom mud. Typically, a depth of 10-12 cm of mud is extracted after the drying process, a task typically carried out through physical labor. The desilting procedure effectively eliminates unwanted silt and bottom mud, which is nutrient-rich, without disrupting subsequent crops. It's essential to transport the removed materials away from the ponds to prevent them from being washed back into the pond during rainfall, rather than spreading them over the dike or bunds. Ensuring a level pond bottom promotes efficient rutting and facilitates fish harvesting.

Dike and canal reconstruction

To prevent fish from escaping during regular flooding, it's crucial to repair and raise the broken pond dyke adequately. Planting grass or other vegetation on the dykes can effectively mitigate erosion during the monsoon months and reduce turbidity issues.

Dike reconstruction is the primary work of pond preparation. It is very essential for the following reasons

To mitigate overflow during the rainy season:

- Prevent dike rupture.
- Shape the pond appropriately.
- Maintain a consistent slope.

Process

Dike reconstruction should be undertaken towards the end of the dry season, preferably between February and April. It's essential to ensure that the inner side of the dike is properly cut or smoothed using mud or soil. The dike should be elevated to a height capable of withstanding floodwaters. Maintaining a slope of 1:2 in relation to the pond's depth is crucial. Additionally, the top of the pond dike should be 3 feet wide. Existing dikes should be repaired by adding new soil, and well- rooted grass should be cultivated along the pond's dike.

Ploughing / Tilling of the Bottom

After removing the bottom silt, if the soil develops cracks, the top 5-8 cm is ploughed diagonally twice, followed by drying for 3-10 days.

Eradication of Aquatic Weeds from Fish Pond

Aquatic weeds are characterized as unwanted vegetation that proliferates excessively in water bodies. They can be categorized based on their habitat into floating, submerged, emergent, marginal, filamentous, and algal blooms.

- **Floating weeds** are characterized by foliage (leaves) that extend above the water's surface, with roots hanging freely. Examples include Eichhornia, Pistia, Spirodela, Lemna, Azolla, Wolffia, Salvinia, and Riccia sp.
- **Emergent weeds** are plants whose leaves and flowers extend above the water's surface but are rooted in the bottom. Examples include Nymphaea, Nelumbo, and Nymphoides.
- **Submerged weeds** are characterized by leaves submerged in the water, with or without roots. Examples include Hydrilla, Ceratophyllum, Myriophyllum, Ottelia, Ludwigia, Sagittaria, Vallisneria, Aponogeton, Cabomba, Limnophila, and Ceratopteris.
- **Rooted submerged plants** include Hydrilla, Najas, Ottelia, Vallisneria, Potamogeton, Lagarosiphon, Chara, and Nitella.
- **Rootless submerged plants** include Ceratophyllum and Utricularia.
- **Marginal weeds** grow in shallow areas near the shore of water bodies. Examples include Typha, Cyperus, Ipomoea, Eleocharis, Cryptocoryne, and Echinodorus.

Aquatic Weeds Scientific Name	Common Name
Eichornia Crasspies	Water Hyacinth
Lemna	Duck Weed
Pistia	Water Lettuce
Azolla	Water fern
Nelumbo	Water Liily
Myriophyllum	Milfoil
Hydrilla	Tape grass
Vallisneria	Eel grass
Typha	Cattail
Ipomea	Water Spinach

Control Measures

Control of Aquatic Weed

Pond preparation is a crucial task, particularly in ensuring optimal productivity. A clean pond, devoid of excessive aquatic weeds, is essential for maximizing production. All types of aquatic vegetation—be they floating, submerged, or emergent—must be removed from the pond. These weeds impede primary productivity by absorbing nutrients from the water and soil, obstructing the normal penetration of sunlight, and hindering netting operations.

It's imperative to ensure ample sunlight reaches the pond surface to foster natural food production and enhance fish yield. While certain aquatic weeds may serve as food sources for certain fish species like grass carp, not all weeds are beneficial for pond ecosystems. Aquatic weeds derive their nourishment from water and soil, posing a threat to plankton reproduction and growth. Excessive weed cover obstructs sunlight penetration, further inhibiting plankton growth. Additionally, these weeds reduce dissolved oxygen levels and impede fish movement.

Methods of Weed Control

- Manual method
- Mechanical method
- Chemical method
- Biological method

Physical Methods

Aquatic weed control involves manual methods such as hand-picking or mechanical techniques utilizing tools like sickles, scythes, blades, nets, wooden rakes, cullers, haulers, and cranes to remove them. Water hyacinth, often termed the worst aquatic plant globally, can be eradicated through manual control methods like hand pulling or raking, as well as mechanical removal employing machinery.

Chemical Methods

To effectively control aquatic weeds, various chemicals are applied at different dosages depending on the season, timing, and the type of infestation present. Herbicides are commonly used for this purpose, with the timing and application method varying based on the specific weed species and habitat. Herbicide application is generally preferred over mechanical methods due to its ease, speed, and cost-effectiveness. However, herbicide use can leave residues in water, posing challenges in areas lacking water use control. Submerged weeds, such as Hydrilla, are particularly challenging to control as they cannot be easily targeted with herbicide sprays or mechanical removal. Commonly used methods include using a combination of herbicides like Quiquat and copper for Hydrilla control, as well as employing techniques like floating booms to concentrate and remove weed masses. Additionally, gibberellin synthesis inhibitors show promise in reducing main stem length, especially under low-light conditions, offering a potential solution for managing certain aquatic weed species like Hydrilla.

Biological Methods

Biological weed control represents an efficient and cost-effective method, where waste vegetative matter is repurposed into protein-rich fish flesh. Various biological agents utilized include fishes, insects, snails, and mammals.

Fishes play a crucial role in biological control of aquatic vegetation, with several herbivorous species known to feed on aquatic weeds. Among them, the grass carp (*Ctenopharyngodon idella*) and silver carp (*Hypophthalmichthys molitrix*) are particularly valued for their ability to consume a wide range of aquatic weeds such as Potamogeton, Lemna, Ceratophyllum, Elodea, Hydrocharis, Vallisneria, and Myriophyllum. Compared to mechanical or chemical methods, fish serve as more sustainable and enduring agents for weed control.

Use of Snails for Biological Control

Significant success has been achieved in the control of aquatic weeds such as Ceratophyllum demersum, Najas guadalupensis, and Potamogeton illinoensis through the utilization of snails. Complete control was observed in these cases. Additionally, partial control was achieved for Pistia stratoites and Alternanthera philoxeroides, while Eichhornia crassipes experienced significant retardation in growth and flowering due to the root pruning action of the snail. Marisa cornuarietis, a snail species known to feed on numerous aquatic plants, shows promising potential for weed control. Moreover, its ability to disrupt the breeding sites of snail vectors for Bilharzia suggests potential applications in non-rice areas. Pomacea australis, another South American snail, is also under consideration for weed control purposes.

Use of Insects for Biological Control

Water Hyacinth (Eichhornia crassipes (Martius) Solms; family: Pontederiaceae) and Water Lettuce (Pistia stratiotes L.; family: Araceae) are troublesome monocotyledonous aquatic weeds that often clog waterways in tropical and subtropical regions worldwide. Water Hyacinth, in particular, has gained notoriety as one of the most invasive aquatic plants globally. However, these weeds can be effectively managed through biological control methods, such as the introduction of insects like weevils.

SI.No	Type of Weed	Chemical and dose	Mode of application
1	Blue green algae andalgal weeds	a) 1 ppm copper sulphate + 2.25 ppm Na.EDTA b) 0.5-1.0 ppm Simazine c) 0.1-0.3 ppm Diuron(1 kg/ha) d) 0.2 ppm Dequat (1 kg/ha) e) 0.2 ppm paraquat(1 kg/ha)	Copper Sulphate packed in cloth bags is immersed in water, applied once in a week; if needed are a number of weeks; other chemicals are sprayed.
2	Floating weeds	a) 80% 2-4 dichlorophenoxy acetic acid (5- 7 kg/ha) b) 80% 2-4 D (4-6 Kg/ha)+small quantity of 0.1% surf detergent c) Simazine(5 Kg/ha) d) 15 ppm ammonia e) 0.2 ppm Dequat(1 kg/ha) f) 0.2 ppm Paraquat(1 kg/ha)	spraying
3	Emergent Weeds	a) 80% 2-4 D (4-6 Kg/ha)+small quantity of 0.1% surf detergent b) 5%Dalapon (2-2 dichloropropionic acid)(5-10 Kg/ha) c) Dichlorophenyl (2-6 dichlorobenzononitrile) (10 Kg/ha) d) Fenae(2,3, 6 trichlorophenyl acetic acid) (10 kg/ha)	Spraying
4	Submerged weeds	a) 15 ppm ammonia b) 3-5 ppm simazine c) 250 ppm urea 4 ppm sodium arsenite	Spraying
5	Marginal weeds	a) 80% 2-4 D (4-6 Kg/ha)+small quantityof 0.1% surf detergent b) 0.2 ppm Dequat(1 kg/ha) c) 0.2 ppm Paraquat(1 kg/ha) d) Dichlorophenyl (2-6 dichlorobenzononitrile) (10 Kg/ha) e) Fenae(2,3, 6 trichlorophenyl acetic acid) (10 kg/ha)	Spraying

Eradication of Aquatic Insects from Fish Culture Pond

Aquatic insects play essential roles in the ecosystem dynamics of freshwater bodies, contributing significantly to organic matter processing, nutrient cycling, and serving as prey for various predators, both aquatic and terrestrial.

A considerable number of aquatic insects, in their larval or adult stages, pose threats to fish hatchlings and fry, competing with them for food resources. Common insect predators include beetles, bugs, and dragonfly nymphs.

Within the beetle family, species like the diving beetle (Cybister), water scavenger beetle (Sternolophus), and whirling beetle (Gyrinus) are particularly notorious for their predatory behavior. Backswimmers (Anisops), often found in swarms in manured ponds during the rainy season, can cause significant damage.

Predatory insects such as the water scorpion (Laccotrephes), giant water bug (Belostoma), and water stick insect (Ranatra) target carp spawn and fry, posing a threat to their survival.

Control of Aquatic Insects From Fish Culture Pond

- The most effective method for controlling aquatic insects is by applying oil soap emulsion on the water surface. Additionally, certain insecticides are utilized for this purpose.
- In nursery pond management, both vegetable and mineral oils are employed. Mustard and linseed oils, for instance, are toxic to young fry and Zooplanktons. Mustard oil, along with other vegetable oils, is emulsified with inexpensive soap at a ratio of 3:1 (e.g., 56 kg mustard oil with 18 kg soap per hectare) 12-24 hours prior to stocking.
- Alternatively, turpentine oil at a rate of 75 liters per hectare, or commercial detergents like "Teepol B306," can be used as substitutes for soap in the emulsion. The recommended dose of Teepol is 560 ml to be emulsified with 56 kg of mustard oil.
- Gammexene at 0.6 to 1 ppm concentration is highly effective, capable of killing insects within 6 hours. Kerosene oil at a rate of 80-100 liters per hectare or diesel oil at 50-60 liters per hectare, mixed with 1/3 to 1/4 washing soap, are also utilized for insect control.

Control of Predators and Undesirable Species

Predators

Predators are species that hunt and consume prey as their primary source of food, such as shoal, boal, chital, and other carnivorous animals.

Undesirable Species

Undesirable species, commonly referred to as weed fishes, are those that are not intentionally cultivated alongside specific or desirable species in a pond. Instead, they naturally proliferate alongside the cultured species.

Reasons for Control

- Predatory fishes consume the fry of cultured species, with predators requiring 10- 12 kg of other fishes for every 1 kg of their own growth.
- Undesirable species compete for the food intended for cultured species.
- Undesirable species can introduce diseases that affect other cultured species.

Eradication of Undesirable Fishes

Prior to stocking the pond with fingerlings of desirable species, it is essential to eliminate all predatory and unwanted fishes. This can be achieved through methods such as complete dewatering of the pond or the use of fish toxicants. Some commonly employed efficient fish toxicants include:

Bleaching Powder

Bleaching Powder (Calcium hypochlorite) can be applied at a rate of 25-30 ppm. The required quantity is dissolved in water and spread over the water surface. After 3-4 hours of spraying, the distressed and deceased fish are removed by netting. The toxicity of the solution lasts for 7-8 days. Any killed fish can be prepared for human consumption.

Rotenone

When applied at a concentration of 2-3 ppm, Rotenone effectively eliminates all fish species except for shrimps. The fish killed by Rotenone can still be consumed by humans. However, its higher cost and limited availability are drawbacks. The toxic effects of Rotenone persist for approximately 10-12 days.

Tea Seed Cake

When applied at a concentration of 2-3 ppm, Rotenone effectively eliminates all fish species except for shrimps. The fish killed by Rotenone can still be consumed by humans. However, its higher cost and limited availability are drawbacks. The toxic effects of Rotenone persist for approximately 10-12 days.

Mahua Oil Cake

Mahua (*Bassia latipolia*) oil cake, with its saponin content ranging from 4-6%, proves lethal to fish when applied at a concentration of 250 ppm. The toxic effects persist for approximately 10-15 days under normal conditions. Fish killed by this method remain suitable for human consumption, and the oil cake also serves as a beneficial organic fertilizer in the pond.

The recommended doses of various fish poisons per hectare or per meter. The dosesinclude:

- **Bleaching powder:** 350 – 500 kg/ha/m
- **Mahua oil cake:** 2,500 kg/ha/m

- **Anhydrous ammonia:** 20 – 30 kg/ha/m
- **Powdered seed of Croton tiglium:** 30 – 50 kg/ha/m
- **Root powder of Milletia pachycarpa:** 40 – 50 kg/ha/m
- **Seed powder of Milletia piecidia:** 40 – 50 kg/ha/m
- **Seed powder of Barringtonia acutanqula:** 150 kg/ha/m
- **Seed meal of tamarind (Tamarindus indica):** 1,750 – 2,000 kg/ha/m
- **Tea seed cake (Camellia sinensis):** 750 kg/ha/m.

Liming

- Liming is a crucial aspect of pond preparation, serving several key objectives, including:
- Maintaining the pH of soil and water above 6
- Enhancing the efficacy of fertilizers
- Reducing water turbidity
- Controlling the decrease of toxic gases
- Promoting a cleaner pond environment
- Boosting productivity levels
- Liming of a fish pond offers numerous advantages, including:
- Neutralizing soil acidity and establishing a buffer system to prevent drastic fluctuations in water pH from acidic to alkaline conditions.
- Eliminating fish pathogens and their intermediate life stages.
- Converting unsuitable acidic water conditions to suitable alkaline conditions.
- Neutralizing undesirable iron compounds present in fish ponds.
- Promoting mineralization of soil, which is beneficial for fish ponds.
- Settling excess dissolved organic matter, reducing the occurrence of oxygen depletion.
- Acting as a determinant and enhancing hygiene levels in the pond.

The buffering action of calcium provided by lime is of paramount importance among its various advantages. Lime serves both prophylactic and therapeutic purposes. It is recommended to apply lime treatment to ponds before the initial manuring process.

Requirements of Lime for Different Types of Pond Soils

The lime requirement for different soil pH levels varies according to acidity or alkalinity:

- Highly acidic soils with a pH of 4.0-4.9 necessitate 2000 kg/ha of lime.
- Moderately acidic soils with a pH of 5.0-6.4 require 1000 kg/ha of lime.
- Nearly neutral soils with a pH of 6.5-7.4 need 500 kg/ha of lime.

- Mildly alkaline soils with a pH of 7.5-8.4 require 200 kg/ha of lime.
- Highly alkaline soils with a pH of 8.5-9.5 do not require lime application.

Water Filling

When preparing a pond for fish stocking, the water filling process is crucial. Initially, before applying manure or fertilizer, it's advisable to maintain a low water depth to maximize the effect of nutrients for natural food production. After 1 to 2 weeks of manuring/fertilizing, the water depth should be raised to the required level before stocking fish seeds. The average water depth in a pond plays a significant role in fish culture and is influenced by factors such as rainfall, evaporation losses, seepage, and water use for irrigation. During the summer, water may need to be supplemented from nearby sources, while during the monsoon, drainage may be necessary to maintain the desired water depth in the pond.

Manuring / Fertilizing

Manuring or fertilizing is a crucial step in pond preparation. Zooplankton and phytoplankton serve as the primary natural food sources for fish.

Objective of Fertilizing

- To enhance the production of natural food.
- To augment the levels of nitrogen, phosphorus, potassium, and other essential nutrients.

Types of Fertilizing

Two main types of fertilization are typically employed: organic and inorganic.

- Organic manures, such as cow dung and poultry manure, are directly applied to foster the development of natural food in the pond.
- Inorganic fertilizers, like urea and superphosphate, are also utilized during pond preparation. Urea is typically applied at a rate of 20-30 kg/ha, dissolved in water, and spread across the pond surface. Similarly, superphosphate is applied at a rate of 10-15 kg/ha.
- In undrainable ponds where water exchange is limited, the physicochemical properties of the water are largely influenced by the bottom soil. Therefore, the fertilizer requirements vary based on soil productivity levels.
- Organic manuring not only supplements nutrients in the water but also aids in improving soil texture. In the absence of organic manures, the use of inorganic fertilizers is recommended.
- A proper schedule for fertilization should be devised based on detailed soil and water analysis data. Fertilizers should only be applied under suitable environmental conditions, including ample sunshine, sufficient oxygen content, and adequate water levels.

- Application of fertilizers is typically done by dissolving them in water and spraying them over the pond surface. Morning hours, between 9-10 o'clock, are considered optimal for fertilization to avoid oxygen depletion due to faster decomposition at night.
- Manuring should be avoided during fully cloudy and rainy days, as well as during algal blooms.
- To optimize fish production, it is advisable to provide the total fertilizer dose in small portions, preferably through daily applications. In integrated livestock-cum- fish culture systems, the excreta of cohabiting animals contribute to the nutrient supply, enhancing fish production.
- Maintaining the appropriate water depth is crucial for fish culture and can be influenced by factors such as rainfall, evaporation, and seepage. Adjustments may be made by adding water from nearby sources during dry spells or draining excess water during the monsoon season to maintain optimal water depth.
- In ponds where low water depth may lead to oxygen depletion due to metabolite accumulation, fish growth can be adversely affected. However, this issue is less common in seasonal ponds.

Periodicity of Fertilizers in Nursery, Rearing, Stocking Pond

Different ponds require varying applications of manure and fertilizers at different intervals:

- In the nursery pond, organic manure is applied every 3 weeks.
- The rearing pond receives both organic manure and inorganic fertilizer on a daily basis, with applications made every 3 weeks.
- For the stocking pond, organic manure and inorganic fertilizer are applied monthly.

Stocking

When transferring 3-4 days old spawns from the hatchery to the nursery tanks, it's preferable to do so during the cooler morning hours. Stocking should occur at a rate of 30-50 lakh ha^{-1}, although higher densities of 10-20 million ha^{-1} can be utilized for nursery rearing in cement tanks with adequate water circulation and quality maintenance. While mono-species rearing is typically recommended, multi-species rearing may be practiced if there is a shortage of ponds. For short-distance transport, open containers filled with sufficient water can be used, while for long-distance transport, polythene bags filled with water and oxygen are most suitable. To acclimatize seeds transported in open containers, pond water is mixed with the container water for temperature adjustment. Seeds transported in closed polythene bags should be floated in pond water before opening, allowing the seeds to swim out on their own. To minimize handling stress, it's important to avoid touching or catching the seeds by hand during stocking.

Chapter-17

Post-Stocking Management in Carp Farming

Post Stocking Pond Management in Carp Farming

This phase encompasses the tasks from the stocking of fingerlings to the final fish harvesting from the pond. These activities include manuring, feeding, monitoring growth and health, assessing water quality, and conducting harvesting operations.

Feeding Management in Nursery Pond

Immediately after stocking, the fish begin to forage on natural pond food regardless of their life cycle stage. Spawn exhibit voracious feeding behavior on plankton, necessitating prompt provision of supplementary feed. In nursery ponds, where spawn are raised for approximately two weeks until the fry stage, supplementary feed in the form of fine powder is scattered daily on the pond surface during the morning hours, following prescribed rates as outlined below –

- For supplementary feeding, a blend of finely powdered groundnut or mustard oil cake and rice bran polish, in equal proportions by weight, is provided to the spawn. Micronutrients like cobalt chloride may also be added to enhance results.
- 'Sushama,' a balanced fish feed produced by the Fisheries Research Centre at Assam Agricultural University, Jorhat, is renowned for its efficacy in this regard.

- Typically, finely ground feed is dispersed by broadcasting over the pond surface starting from the day of stocking during morning hours, with feeding ceasing one day before the scheduled harvesting.
- Under adverse ecological conditions, feeding should be temporarily halted. In grass carp spawn rearing programs, chopped and minced vegetation such as Wolffia and Hydrilla may be offered as feed after 10- 12 days of stocking.

Harvesting of Fry

- Fry typically reach a size of 25-30 mm within approximately 15 days. They are harvested using fine-meshed (1.5 mm) dragnets or nursery nets during the cool morning hours, avoiding heavy showers or excessively hot days.
- In Assam's agro-climatic conditions, slower growth rates are observed, particularly during April-May when temperatures are not optimal. Consequently, it may take around 20-25 days to reach the fry stage.
- Fry produced in nursery ponds are harvested once they attain the desired length and are then transferred to rearing ponds to be raised to the fingerling stage, which is ideal for stocking in culture or grow-out ponds.
- In a seed-raising farm, a nursery pond can accommodate the rearing of 5-6 batches of spawn to the fry stage, with an average culture period of 20-25 days per batch.

Feeding Management in Rearing Pond

After stocking, supplementary feeding is initiated following these guidelines:

- A mixture of rice bran and mustard oil cake, in a 1:1 ratio by weight, is commonly used for preparing supplementary feed for fry. Alternatively, 'Sushama,' the AAU fish feed, may be employed for enhanced results. The quantity of supplementary feed applied varies based on the body weight and age of the fry.
- During the first month of rearing, supplementary feed is provided at a rate of 8-10% of the total body weight of the stocked biomass. In the second and third months, feed rates decrease to 6-8% and 4-6% of their body weight, respectively.
- Both bag or tray feeding and broadcasting methods are effective for feeding the fry.
- Herbivorous fishes like grass carp benefit from aquatic vegetation such as Wolfia, Lemna, Spirodela, Azolla, Hydrilla, Valisneria, Najas, Potamogeton, and selected terrestrial or semi-aquatic vegetation. These should be provided after thorough washing and mincing.
- Feeding is typically withheld during cloudy days and when there is algal bloom in the pond.

Harvesting of Fingerling

Fingerling harvesting is best conducted in the morning hours using repeated netting with dragnets of appropriate mesh size. Supplementary feeding should cease one day before the scheduled harvesting day. Utilizing rearing ponds, it's feasible to rear 2-3 batches of fingerlings in a season, with each batch requiringan average culture period of 2-3 months.

Feeding Management in Stocking Ponds

- Supplementary feeding, consisting of a 1:1 mixture of any oil cake (Mustard/Groundnut/Til/Coconut) and rice polish or rice/wheat bran, is administered at a rate of 3% of the stocked fish's body weight per day, starting from the second day of stocking. The quantity can be adjusted based on daily observations of fish consumption rates.
- Feeding fish twice daily, in the morning and evening, enhances feed utilization.
- Mustard oil cake and rice bran are mixed to form dough, which is then placed in bamboo trays and positioned at a specific depth in the pond. These feeding trays are tethered to poles fixed on the embankment to prevent displacement, with 10 to 15 trays typically sufficient for a 1 ha fish pond. Adjustments to the depth of the feeding trays are made based on the pond's depth by altering the length of the rope.
- Regular health checks of fish are essential, requiring periodic netting at least once every two months. Healthy fish exhibit robust growth and vigorous feeding behavior. If feed consumption is inadequate, it could indicate fish disease or deteriorating water quality. Monitoring water quality, pond hygiene, and fish health is crucial for implementing remedial actions as needed.

Harvesting of Fish

Typically, fish in stocking ponds are given a growth period of 12 months, during which Catla reaches a weight of 800 gm – 1 kg; Rohu 600 – 800 gm; Mrigal 400 – 600 gm; Silver carp 1.0 – 1.2 kg; Grass carp 1.0 – 1.5 kg, and Common carp 800 gm – 1.0 kg.

Chapter-18

Criteria for Selection of Candidate Species for Aquaculture

Rate of Growth

Fishes that exhibit rapid growth and achieve significant size within a short timeframe are highly favorable for cultivation, such as Indian major Carps.

Short Feeding Chain

Fishes with a short feeding chain are considered optimal as they minimize energy loss between trophic levels. For instance, species feeding on detritus, plankton, or vegetation offer additional advantages due to their tolerance of other species within a pond.

Adaptation to Climate

This condition is crucial as it restricts the utilization of both cold-water and warm-water species. Cold-water species such as Salmonids are unable to withstand warm water temperatures, while warm-water species like Tilapia spp. and Indian major carps struggle to endure the cold climates typical of temperate regions.

Consumer Liking

It is imperative to consider consumer preferences when selecting a species for aquaculture. For example, Silver carp may not be favored due to factors such as its spines and lower shelf life.

Aptitude for Artificial Food

To achieve a high production rate, it's imperative to rear fish species that readily accept artificial feed.

Tolerance to Fluctuations in Physic-Chemical Conditions of Water

This is a highly desirable characteristic of a species chosen for aquaculture. Such fish can endure handling and transportation stress effectively.

Resistance to Common Fish Diseases and Parasites

Lernea infection tends to affect Catla more severely than other fish species, while Rainbow Trout (RBT) demonstrates greater resistance to the IPN virus compared to Brown Trout (BT).

Easy Reproduction Under Controlled Conditions

To maintain a consistent and readily available supply of fish seed for rearing, it's advantageous for cultured fish to reproduce in captivity. However, if reproduction becomes overly prolific, it can lead to population density issues.

Amiability to Live Together

The ability to coexist without disturbing other species is particularly important for fishes utilized in polyculture systems. In this regard, it is advisable to culture carnivorous fishes separately.

More Edible Flesh Per Unit Weight

Species that yield higher flesh-to-weight ratios are more cost-effective.

Chapter-19

Major Candidate Species for Freshwater Aquaculture

Catla *(Catla catla)*

- Catla, the swiftest-growing Indian major carp, is a crucial element of the carp polyculture system in the country. It reaches maturity in its second year and exclusively breeds in flowing rivers and streams during the monsoon season (May to September).

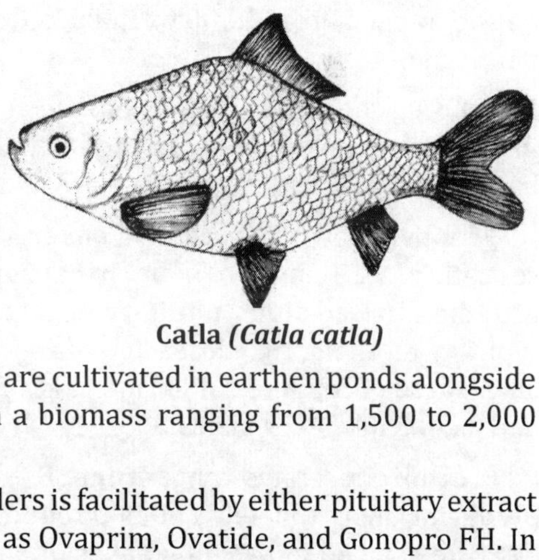

Catla *(Catla catla)*

- Two to three-year-old brooders are cultivated in earthen ponds alongside other Indian major carps, with a biomass ranging from 1,500 to 2,000 kg/ha.
- Induced breeding of these brooders is facilitated by either pituitary extract or commercial hormones, such as Ovaprim, Ovatide, and Gonopro FH. In the case of pituitary extract, females receive two doses of 2-3 mg/kg and 4-6 mg/kg, while males receive a single dose of 2-3 mg/kg.
- Conversely, commercial hormones are administered as a single dose of 0.4-0.5 ml/kg to females and 0.2-0.3 ml/kg to males. In current practices of induced breeding within circular echo-hatcheries, a 1:1 ratio of females to males is maintained.

- Following injection, the induced broods are released into breeding pools, with spawning typically occurring within 6- 8 hours. On average, 1.0-1.5 lakh eggs/kg female are produced, with fertilization rates exceeding 90%. These eggs are transferred to hatching tanks, where they are maintained in a floating condition with appropriate water circulation.
- Hatching typically occurs within 18-22 hours at temperatures of 27-28°C, with a success rate of 80-95%. After hatching, the hatchlings remain in the hatching tank for an additional two days without feeding, as they utilize their yolk-sac reserves, before being transferred to the nursery.
- The spawn are then reared in earthen nursery ponds and concrete tanks at stocking densities of 3-5 million/ha and 8-10 million/ha, respectively. Over a period of 15-20 days, they grow to an average fry size of 25 mm, with survival rates ranging from 30-35% to 50-60%.

Rohu *(Labeo rohita)*

Rohu, the predominant Indian major carp in the Indian Sub-continent, reaches maturity in its second year and reproduces exclusively in riverine conditions during the monsoon season.

Labeo rohita **(Rohu)**

For captive breeding purposes, brooders of the 2+ year class are cultivated in earthen ponds at a biomass density ranging from 1,500 to 2,000 kg per hectare.

During the spawning season, the sexes can be distinguished, with females displaying a swollen abdomen and soft pectoral fins, while males possess a slender body, rough pectoral fins, and release milky-white milt when pressed on the belly.

Captive spawning is carried out either in hapas or hatcheries, maintaining a sex ratio of 1:2 (female to male) in hapas and 1:1 in circular hatcheries. Brooders are administered either pituitary extract or commercial hormones, with females typically receiving two doses and males one dose of pituitary extract, whereas a single dose of commercial hormone is administered at 0.4-0.5 ml/kg for females and 0.2-0.3 ml/kg for males.

Fertilization rates range from 80% to 95%, with each kilogram of female producing approximately 200,000 round, transparent, semi-buoyant eggs. These eggs are hatched in hatching hapas or spawning tanks within hatcheries, taking approximately 8- 22 hours at a temperature of 27-28°C.

After hatching, the larvae are kept in hatching tanks for an additional two days before being transferred to nurseries for further rearing. Nursery rearing is conducted in earthen ponds at stocking densities ranging from 5 to 10 million per hectare, with higher densities of 10 to 20 million per hectare possible in concrete nurseries.

This breeding and seed production technology is widely practiced across all Indian states.

Mrigal *(Cirhinus mrigala)*

- An important carp species, highly favored by consumers due to its moderate growth rate, is cultivated alongside Indian major carps in a polyculture system, typically constituting a minimal proportion of approximately 5%.

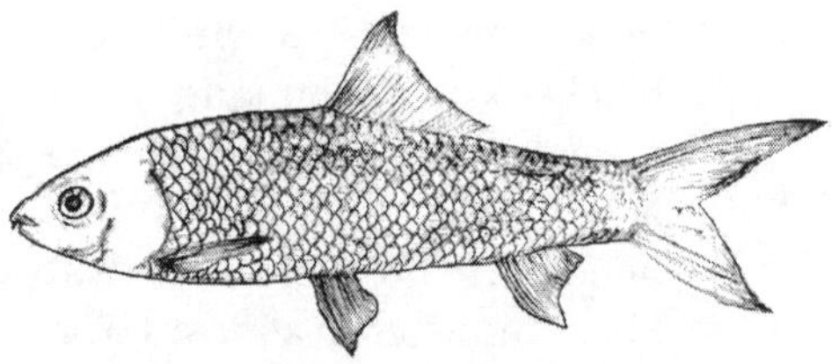

***(Cirhinus mrigala)* Mrigal**

- This fish reaches maturity within two years under pond conditions, with the breeding season typically spanning from May to July. Brooders are nurtured within the ponds alongside other major carps.
- In induced breeding practices, selected brooders are placed in spawning tanks within the hatchery, maintaining a female- to-male ratio of 1:2. Breeding is facilitated through a single injection of hormonal formulations such as Ovaprim, Ovatide, or Gonopro FH, administered at 0.4-0.5 ml/kg for females and 0.2-0.3 ml/kg for males.
- Spawning typically occurs within 6-8 hours, with females exhibiting a fecundity ranging from 1.5 to 2.0 lakh eggs per kilogram of body weight. The transparent, demersal eggs possess a diameter of 3.0-3.5 mm.
- Following hydration, the eggs are transferred to hatching tanks, where they are maintained in a floating state with gentle water circulation. Hatching usually transpires within 15-18 hours, after which the hatchlings are retained under similar conditions for an additional two days to absorb their yolk-sacs, providing essential nutrition during this incubation period. Subsequently, the spawn is deemed ready for feeding and is transferred to the nursery.
- The spawn is reared to an average fry size of approximately 20 mm over a period of 15-20 days, with stocking densities ranging from 5 to 10 million per hectare.
- Survival rates in earthen nursery ponds and concrete tanks typically range from 30% to 50%. Recognized as the state fish of Haryana, its seed production is widely practiced across several states.

Kalbasu *(Labeo calbasu)*

- A sought-after carp species with moderate growth rates is cultivated alongside Indian major carps in a polyculture system, typically representing a modest proportion of approximately 5%. Maturing within two years under pond conditions, the breeding season for this species typically spans from May to July, with brooders raised alongside other major carps within the ponds.

- For induced breeding, carefully selected brooders are housed in spawning tanks within the hatchery, maintaining a female-to-male ratio of 1:2. Breeding is initiated through a single injection of hormonal formulations such as Ovaprim, Ovatide, or Gonopro FH, administered at doses of 0.4-0.5 ml/kg for females and 0.2-0.3 ml/kg for males.

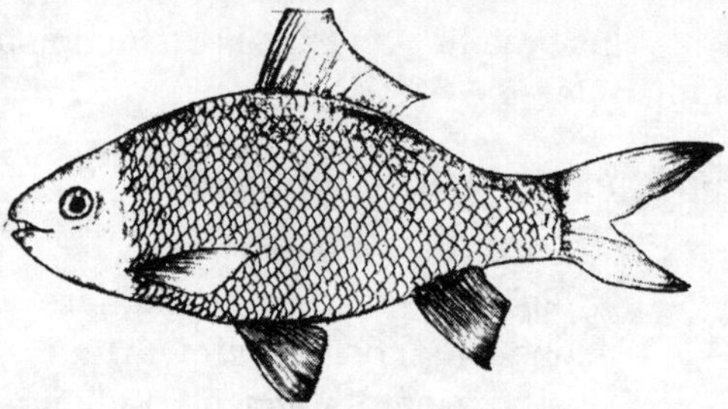

***Labeo calbasu* (Kalbasu)**

- Spawning typically occurs within 6-8 hours, with females producing a fecundity ranging from 1.5 to 2.0 lakh eggs per kilogram of body weight. The eggs, transparent and demersal with a diameter of 3.0-3.5 mm, are then transferred to hatching tanks, where they are kept afloat with mild water circulation. Hatching usually takes place within 15-18 hours, after which the hatchlings are retained under similar conditions for an additional two days to absorb their yolk-sacs, providing essential nutrients during this incubation period.
- Following the absorption of the yolk-sacs within 72 hours, the spawn is ready for feeding and subsequently transferred to the nursery. Here, the spawn is reared to an average fry size of approximately 20 mm over a period of 15-20 days, with stocking densities ranging from 5 to 10 million per hectare. Survival rates in earthen nursery ponds and concrete tanks typically range from 30% to 50%. Recognized as the state fish of Haryana, its seed production is widely practiced across several states.

Kuria Rohu *(Labeo gonius)*

- Kuria rohu, a minor carp, holds significant popularity in the north-eastern states of India. It constitutes a smaller portion within certain regions of eastern and north-eastern states' carp polyculture systems.

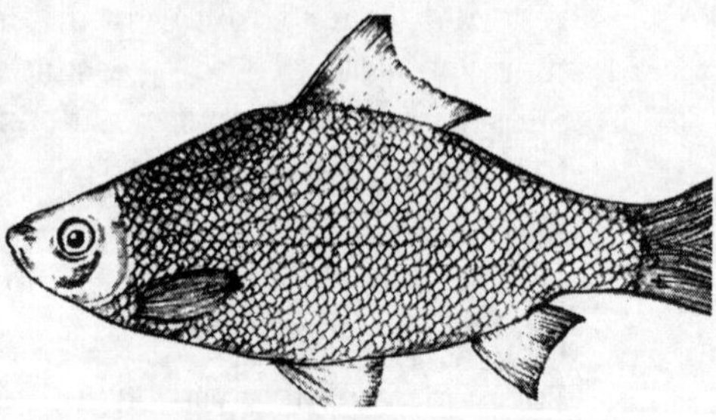

***Labeo gonius* (Kuria Rohu)**

- Maturing within two years, this species breeds during May-August in flowing rivers, with brooders raised alongside Indian major carps in earthen ponds.
- At maturity, sexes are easily distinguishable: females exhibit a bulged abdomen and smooth pectoral fins, while males are slender with rough pectoral fins, releasing milt upon belly pressure.

- Induced breeding of Kuria rohu is achieved through pituitary extract or commercial hormonal formulations like Ovaprim, Ovatide, or Gonopro FH. A single injection of these hormones, at a rate of 0.4 ml/kg for females and 0.2 ml/kg for males, has proven highly successful.
- Selected brooders, maintained at a 1:1 female-to-male ratio post-injection, are transferred to spawning tanks where spawning occurs within 10-11 hours, yielding approximately 1.0-1.5 lakh eggs per kilogram of female weight.
- The fertilized eggs are then moved to hatching tanks, where hatching typically occurs within 18-20 hours. The spawn stage is reached within 50-55 hours, boasting a survival rate of 70-80%.
- Subsequently, the spawn is reared in concrete tanks at densities ranging from 10 to 15 million per hectare for 25-30 days to attain fry of 20-25 mm size, with survival rates ranging from 30% to 40%. Alternatively, rearing in earthen nurseries can be conducted similarly to major carps, albeit at a lower density of 5 million per hectare.
- Seed production of Kuria rohu is particularly practiced in regions such as Assam, West Bengal, and Odisha.

Fringe Lipped Carp *(Labeo fimbriatus)*

- The Fringed-lipped carp, a significant minor carp, is well-suited for polyculture alongside Indian major carps, primarily found in Peninsular India. It achieves maturity within two years in pond conditions but does not breed in captivity, preferring flowing waters like other major carp species.

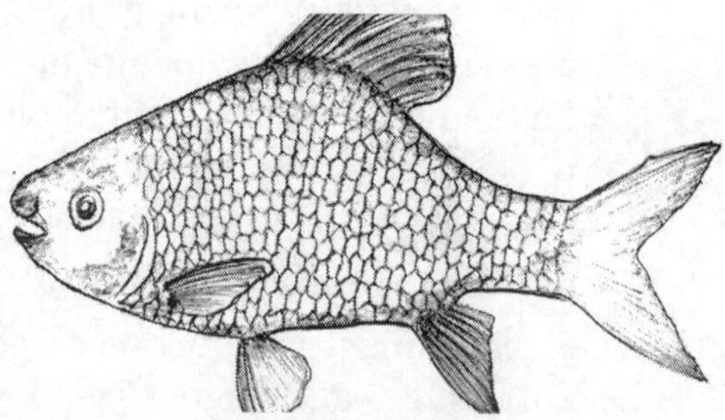

Labeo fimbriatus
(Fringe Lipped Carp)

- At maturity, both males and females display distinguishable characteristics, with females exhibiting smooth pectoral fins and a bulged abdomen, while males possess a slender body with rough pectoral fins.
- For captive breeding, brooders are cultivated in ponds alongside other carp species. Mature pairs, selected at a ratio of 1:2 females to males, are injected with a single dose of commercial hormone (Ovaprim, Ovatide, Gonopro FH) at rates of 0.30 ml/kg for females and 0.15 ml/kg for males. They are then released into the spawning pool at a sex ratio of 1:1 for breeding.
- Spawning typically occurs 6-8 hours after injection, with the hydrated eggs collected and transferred to the hatching pool. These pools are equipped with mild water circulation to maintain the eggs in a buoyant condition. Hatching occurs within 14-16 hours, with hatchlings reaching spawn size

within 50-55 hours. Spawn yield typically ranges from 0.85-

- 1.00 lakh eggs per kilogram of female weight.
- The spawn is reared at stocking densities of 10-15 million per hectare in concrete nursery systems for 20-25 days, achieving a survival rate of 40-50%. Alternatively, in earthen nurseries, stocking is carried out at a lower density of 5 million per hectare.
- Breeding and seed production of the Fringed-lipped carp are primarily practiced in Karnataka, Odisha, and Andhra Pradesh.

Malabar labeo (*Labeo dussumeieri*)

- Malabar Labeo, a cherished endemic carp species found in the Western Ghats, extends its distribution up to North Canara in India and potentially to Sri Lanka, though it is categorized as an endangered fish.

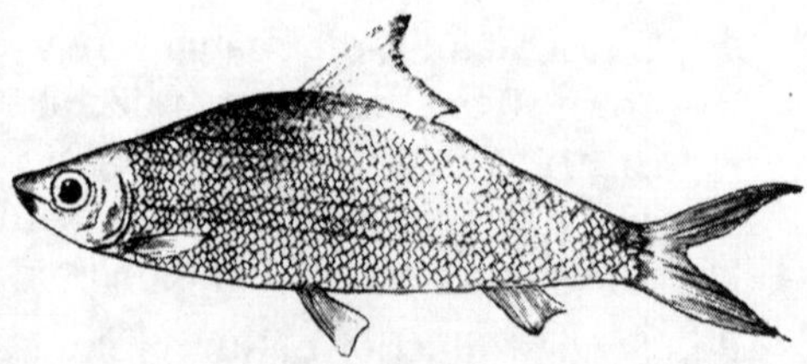

Labeo dussumeieri
(Malabar labeo)

- Mature females are characterized by a bulged abdomen, with eggs extruding from the vent upon gentle pressure. Conversely, mature males exhibit rough surfaces on their pectoral fins and release milky milt upon abdominal pressure.
- In captive conditions, induction breeding occurs during the monsoon season. Fully mature brooders, in a sex ratio of 1:2 females to males, are placed in the spawning tank of a circular carp hatchery for breeding.
- Females undergo induction breeding with pituitary extract, receiving two doses at 2-3 and 4-6 mg/kg body weight, or a single dose of commercial hormone at 0.3 ml/kg. Males receive a single injection of 2-3 mg of pituitary extract/kg or
- 0.2 ml of commercial hormone/kg body weight.
- Spawning takes place within 9-10 hours, yielding spherical, translucent, demersal, nonadhesive fertilized eggs with diameters ranging from 4.0 to 4.5 mm.
- These fertilized eggs hatch within 10-12 hours, boasting hatching rates of 80-95% in the hatching tank, and reach spawn size within two days.
- Subsequently, the spawn is further reared in tanks to raise fry. Seed production of Malabar Labeo is predominantly practiced in the state of Kerala.

Silver Carp *(Hypothalmichthys molitrix)*

- In India, fish typically reach maturity at the age of two years. Brooders belonging to the 2+ year-class are cultivated in earthen ponds alongside other carps, and induced spawning is carried out during the monsoon season.

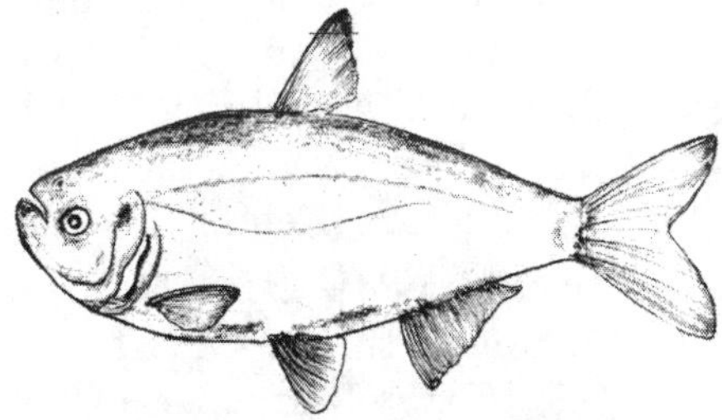

Hypothalmichthys molitrix (Silver Carp)

- Sexual dimorphism becomes evident during the spawning season, with females displaying a bulged abdomen and soft pectoral fins, while males are slender, with rough pectoral fins and release milky white milt from the vent.
- Captive spawning is conducted either in hapas or hatcheries, maintaining a recommended sex ratio of 1:1 in terms of both number and weight. Brooders are induced to spawn by injecting either pituitary extract (PE) or commercially available synthetic gonadotropin hormones (CH) such as Ovaprim, Ovatide, Ovulin, Gonopro, or WOVA-FH.
- For PE, females receive two doses, with the first dose administered at 2-3 mg/kg and the second at 6-8 mg/kg body weight, given at an interval of 6-8 hours. Males are given a single dose of 2-3 mg/kg. In the case of CH, females are administered a single dose of 0.5-0.6 ml/kg, while males receive approximately half of that dose (0.2-0.3 ml/kg) administered to females.
- After 6-8 hours of hormone administration, eggs and spermatozoa are stripped into a dry bowl/tray and mixed for fertilization. Typically, one kilogram of female lays around 0.8-1.0 lakh eggs, with a fertilization rate averaging 50- 60%. The eggs are then hatched either in hatching hapas or incubation chambers within hatcheries.
- The yolk sac of the hatchling is absorbed within 65-76 hours at a water temperature of 28-30ºC, resulting in the development of free-swimming spawn with a hatching rate exceeding 70% and a size of 5-6 mm.
- Spawn is further reared to fry size (20-25 mm) at a density of 500-1000/ m^2 in prepared earthen nurseries for 15-20 days, achieving a survival rate of 30-35%. Alternatively, in concrete tanks, stocking at a density of 1000-2000 spawn/m^3 results in better survival rates ranging from 50-80%.
- This breeding technology has been widely embraced across all states of the country.

Grass carp *(Ctenopharyngodon idella)*

- Grass carp, native to Asia and spanning from China to Eastern Siberia, is globally cultured due to its compatibility in polyculture systems.
- In captivity, breeding is achieved through artificial induction, with brooders of the 2+ year class raised in earthen ponds at a biomass of 1500-2000 kg/ha.

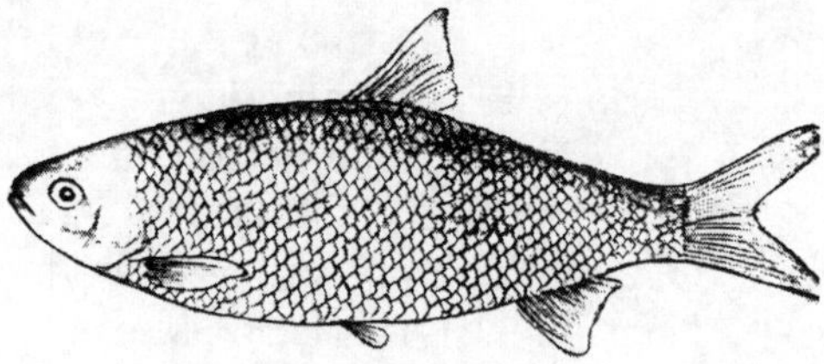

Ctenopharyngodon idella
(Grass carp)

- During the spawning season, sexes exhibit distinct characteristics, with females displaying a bulged abdomen, fleshy genital pore, and soft pectoral fins, while males are slender, with rough pectoral fins and release milky-white milt from the vent upon abdominal pressure.
- Captive spawning occurs in either hapas or circular hatcheries, maintaining a sex ratio of 1:1. Brooders are induced through injections of either pituitary extract or commercially available synthetic Gonadotropin hormones (CH).
- For pituitary extract, females receive two doses: the first at 2-3 mg/kg and the second at 6-8 mg/kg body weight, with males given a single dose of 2-3 mg/kg at the time of the second female dose. In the case of CH, females are administered a single dose of 0.5-0.6 ml/kg, while males receive 0.2-0.3 ml/kg.
- Stripping is performed in hapas, while grass carp breeds naturally in spawning tanks within circular hatcheries. Fecundity of this species is approximately 1.5 lakh/kg body weight of females, with a typical fertilization rate of 90- 95% under ideal conditions.
- Eggs hatch within 18-22 hours at a water temperature of 28-30ºC, and the yolk sac is absorbed within 72 hours, resulting in the development of free-swimming spawn measuring 5-6 mm in size, with a hatching rate of 70%.
- Spawn is reared to fry size (20-25 mm) at a density of 500-1000/m2 in earthen nurseries for 15-20 days, achieving a survival rate of 30-35%. Alternatively, in concrete tanks at a stocking density of 1000-2000 spawn/m3, survival rates are enhanced, ranging from 50-80%.
- This breeding technology is widely practiced across all states of the country

Magur *(Clarius magur)*

- Magur, widely distributed across the Indian sub-continent, is a popular air-breathing catfish known for its high food value and medicinal importance, making it a suitable species for aquaculture diversification.

- While breeding occurs in captive conditions, sourcing seed from the wild proves challenging.
- Captive broodstock is reared in concrete tanks or earthen ponds at a density of 2-3 individuals per square meter, with intermittent water exchange and a diet consisting of compound feed containing 30-35% protein.

Clarius magur
(Magur)

- Matured brooders, typically weighing 100-150 g, are easily identified by their pointed and round genital papilla for males and females, respectively. Females receive injections of Ovaprim, Ovatide, Wova-FH, or GonoproFH at a rate of 1.0-1.5 ml/kg body weight.
- After 16-17 hours post-injection, females are stripped, and their eggs are fertilized with sperm suspension obtained from the dissected testes of males prior to stripping. A female weighing 100 g typically yields 4,000-5,000 eggs.
- The round, dark brown, adhesive eggs undergo 3-4 washings after stripping before being transferred to hatchery tubs/tanks.
- Hatching occurs within 24-26 hours, with the yolk sac absorbed within 70-80 hours.
- Spawn is reared in indoor tanks at a density of 1,000-1,500/m2 for approximately two weeks. Larvae are initially fed with live feed, followed by compound larval feed, to achieve a survival rate of 80% at the fry stage.
- The breeding and seed production technology of Magur has been widely adopted in several states of the country, particularly in West Bengal, Odisha, Assam, and other northeastern states.

Pangasius Pangasius (Pangus catfish*)*

- Pangas, a sizable and popular catfish in the Indian sub-continent, thrives in the large rivers of North, Central, and Southern India, presenting a promising species for diversification in freshwater aquaculture.
- Breeding occurs naturally in flowing waters during the

Pangasius Pangasius
(Pangus catfish)

monsoon season, while brooders are raised in earthen ponds alongside other fish species.

- Sexual dimorphism is evident at maturity, with females displaying a bulged abdomen and males releasing white milt upon abdominal pressure.
- Females receive injections of synthetic hormones such as Ovatide or Ovaprim at a rate of 1.0-1.5 ml/kg body weight to facilitate successful stripping after 13-14 hours of injection.
- Stripping is done using the dry method, and eggs are fertilized with collected milt. Females typically lay 1.2-1.4 lakh eggs/kg body weight, with the adhesive nature of the eggs necessitating the removal of the sticky layer before transferring them to hatching tanks.
- Hatching occurs within a day, yielding hatchlings of 3-4 mm in size, which are subsequently reared in separate tanks within the hatchery at a low stocking density of 500-1,000/m2 to minimize mortality due to cannibalism.
- Larvae are fed with small zooplankton obtained through repeated sieving or Artemia nauplii or tubifex worms starting from the fourth day onwards. Within 15-20 days, the larvae reach an average weight of 30-40 mg, referred to as early fry.
- Early fry are then further stocked in nursery ponds or concrete tanks at a density of 10-15 individuals per square meter and fed with crumbled feed daily to reach a size of 2-3 g within 1-2 months.
- The breeding technology for Pangas is currently being adopted in the state of Odisha.

Freshwater Shark *(Wallago attu)*

- The largest freshwater catfish in India, this species is distributed from Pakistan to Vietnam and Indonesia. It matures within 2 years and spawns between May and September in rivers and large reservoirs.
- For captive breeding, broodstock are raised in ponds and fed with small-sized fishes.

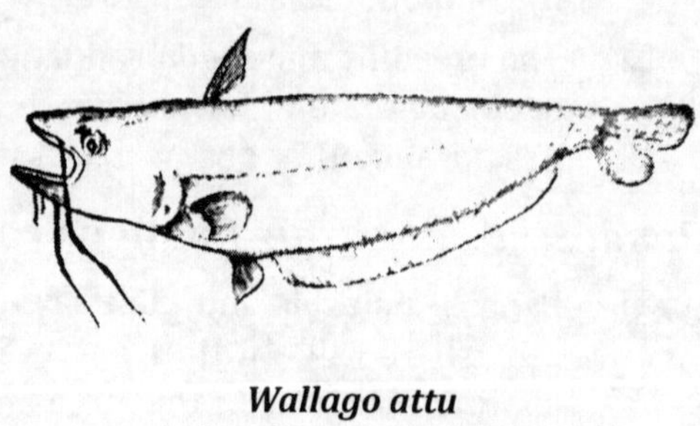

***Wallago attu* (Freshwater Shark)**

- Sexes are separate and distinct, with males possessing a slender body, large serrated pectoral fins, and a swollen pointed genital papilla, while females exhibit a bulging abdomen, soft pectoral spine, and blunt genital papilla.
- Pairing ratios range from 1:1 to 1:2, depending on the size of the fish. Induction is achieved using pituitary gland extract or commercial fish

hormone Ovaprim/GonoproFH. Females injected with pituitary extract receive two doses (4 mg/kg and 12 mg/kg), while males are administered a single dose of 2-3 mg/kg. With commercial hormone, females receive a single dose of 0.5-0.6 ml/kg and males receive 0.3 ml/kg.

- Injected brooders are transferred to spawning tanks equipped with wooden logs and macrophytes such as Hydrilla verticellata. Spawning occurs 6-8 hours post-hormone injection.
- Eggs, which are sticky, adhere to weeds, wooden logs, and tank walls, with a fertilization rate ranging from 70-90%. They are spherical, pale yellow, with a diameter of 2.2-2.8 mm, and fecundity averages around 30,000-40,000 eggs/kg.
- Hatching occurs within 14-18 hours at 28±1°C. Hatchlings begin crawling over the tank bottom on the second day and feed on their siblings, necessitating feeding with zooplankton and cut pieces of insect larvae from the second day onwards.
- Yolk-sac absorption takes place on the fourth day. Spawn is collected and transferred to another tank for growth to the fry stage, where they are fed zooplankton and meshed insects and their larvae. Survival rates at the fry stage are typically low, ranging from 10-20%.
- Successful breeding of this species has been reported in Odisha and Uttar Pradesh.

Striped Murrel *(Channa striatus)*

- Striped murrel, a significant air-breathing food fish, is widely distributed throughout the Indian subcontinent and Southeast Asian countries. Mature brooders can be observed from May to October, although in Southern India, they can be found throughout the year.

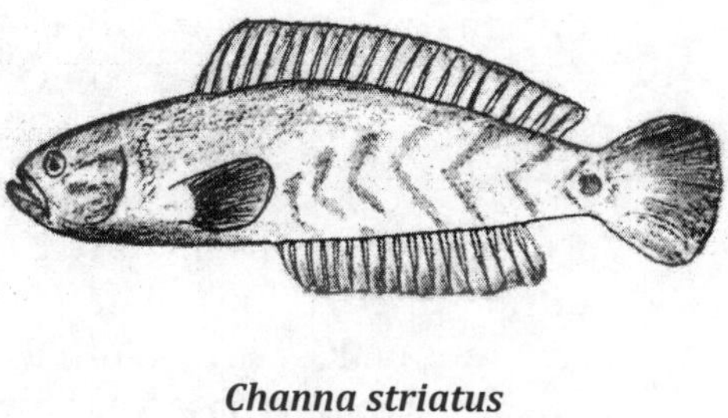

Channa striatus
(Striped Murrel)

- Distinguishing between the sexes requires careful examination to assess maturity. Females display a slightly bulged abdomen and a round-reddish vent, while males exhibit a pale vent and a small, slender anal papilla without oozing milt upon belly pressure.
- Female and male brooders (0.4-1.0 kg) are injected with either HCG at rates of 2000 and 1500 IU/kg body weight, carp pituitary gland extract (PGE) at rates of 20-30 and 10-20 mg/kg body weight, or Ovatide/Ovaprim at rates of 0.6 and 0.4 ml/kg body weight, respectively.

- Spawning occurs within 16-18 hours at temperatures of 26-28°C. The eggs are spherical, transparent, non-adhesive, free-floating, and straw-yellow in color. Generally, 10,000-15,000 eggs/kg female body weight can be obtained at once, with fertilization rates ranging from 70-95%. The fish can spawn 3-4 times a year.
- Hatching takes place within 22-24 hours with a success rate of 70-95%, and the yolk sac is absorbed within 72 hours. The spawn are reared for 7-10 days in small tanks and fed ad libitum with either zooplankton or Artemia nauplii. With proper management measures, a survival rate of 50-60% can be achieved.
- Fry rearing is carried out in outdoor tanks, where they are fed aquatic insects, tubifids, and chopped earthworms mixed with powdered fish meal and soya flour (3:1) at 5-10% of their body weight. Approximately 2-5% of the fry develop into shoot fry, necessitating weekly grading to remove smaller individuals. An average survival rate of 25-30% is achieved with regular grading of shooters.
- The breeding and seed production technology is practiced in West Bengal, Odisha, Telangana, Andhra Pradesh, Tamil Nadu, Uttar Pradesh, and Chhattisgarh.

Giant Murrel *(Channa marulius)*

- The giant murrel, also known as the giant snakehead, holds significant value as an air-breathing food fish and is distributed from the Indian subcontinent to various countries across Southeast Asia.

Channa marulius
(Giant Murrel)

- Broodstocks are cultivated either in earthen ponds or concrete tanks, with matured brooders typically observed between March and October, though they are present year-round in Southern India.
- Sexes are distinct but require careful examination to assess maturity. Upon maturity, females display a slightly bulged abdomen and a round-reddish vent, while males exhibit a pale vent and a small, slender anal papilla without milt discharge upon belly pressure.
- Fully gravid males and females (0.7-1.5 kg) are administered an initial injection of human chorionic gonadotropin (HCG) at 2000 IU/kg body weight, followed by a second dose of HCG at 1500-2000 IU/kg body weight after a 24- hour interval for both sexes.

- Spawning occurs 16-18 hours after the second injection, with females laying 8,000-12,000 eggs/kg. The eggs are spherical, light yellow-transparent, non-adhesive, and free-floating.
- Fertilization rates are typically high, around 85-90%, with hatching taking place within 22-24 hours and achieving a success rate of 80-85%. The newly hatched larvae are transparent and measure 3.4-4.0 mm in total length.
- Larvae are reared in small tanks and fed ad libitum with either zooplankton or Artemia nauplii. With proper management practices, fry survival rates of 50-60% can be achieved.
- Fry rearing is conducted in outdoor tanks, with feeding consisting of aquatic insects, tubifids, and chopped earthworms mixed with powdered fish meal and soya flour (3:1) at 5-10% of their body weight. Regular grading of shooters enables an average survival rate of 25-30%.
- The breeding and seed production technology has been expanded to West Bengal, Odisha, Telangana, and Uttar Pradesh.

Common Carp *(Cyprinus carpio)*

- Originating primarily from Europe and Asia, the common carp stands as the most extensively cultured species worldwide.
- Attaining maturity within its first year, the common carp breeds in confined waters throughout the year, peaking in July-August and March-April.

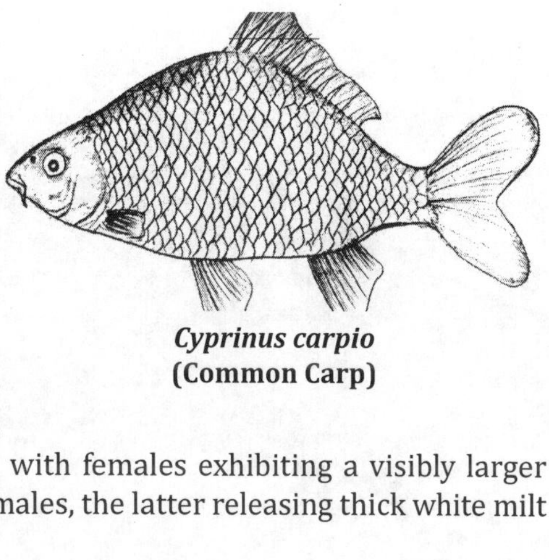

Cyprinus carpio
(Common Carp)

- During the spawning season, sexes are easily distinguished, with females exhibiting a visibly larger bulged abdomen compared to males, the latter releasing thick white milt when their belly is pressed.
- Captive spawning can be conducted in both hapas and hatcheries, maintaining a 1:1 female to male ratio in both setups.
- Brooders are induced through mild doses of either pituitary extract (PE) or commercial hormone (CH), with females receiving a single dose of PG extract at 2-3 mg or CH at 0.1-0.2 ml/kg, and males injected with PE at 2 mg or CH at
- 0.1 ml/kg. The induced brooders are then released into spawning pools or hapas in a 1:1 ratio. To counter the stickiness of common carp eggs, either aquatic weeds or polythene/netting strips are spread on the substratum of the spawning tanks.

- Spawning occurs within 6-8 hours of hormone administration, with one kilogram of female carp laying around 1.5 lakh eggs and a fertilization rate of generally 80-95%.
- The stripping method of spawning involves collecting male and female gametes through stripping and fertilizing them by mixing. To remove the stickiness of the eggs for proper handling and successful hatching, either salt carbamide (fertilization solution) or cream milk method is employed.
- Eggs are hatched either in hatching hapas or spawning tanks within hatcheries, hatching within 36-48 hours. The yolk sac of the hatchlings is absorbed within 72 hours, resulting in free-swimming spawn with a hatching rate of over 70%.
- Spawn are subsequently reared in nursery ponds or concrete tanks for 15 days to raise fry.

Chapter-20

Major Candidate Species for Mariculture

Cobia (*Rachycendron canadum*)

- Cobia, a prized marine migratory pelagic finfish, thrives in tropical, subtropical, and warm-temperate waters worldwide, making it a significant species for mariculture.

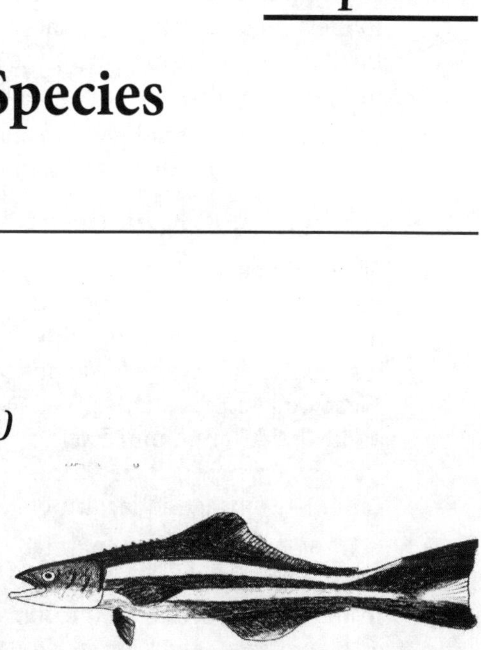

Cobia
(*Rachycendron canadum*)

- Sub-adult cobia, typically weighing 5-6 kg, are captured from the wild and developed into broodstock within sea cages. Gravid males and females are identified through cannulation and tagging, as sexual dimorphism is not readily apparent in this species.
- Mature cobia, weighing approximately 8-10 kg or more, are selected for spawning, with females exhibiting over 20% of intra-ovarian egg diameter exceeding 700 µm. They are then transferred to spawning tanks, which have a capacity exceeding 60 tonnes, maintaining a female to male ratio of 1:2.
- Spawning induction is initiated through a single injection of HCG at 500 IU/kg body weight for females and 250 IU/kg body weight for males. The

fish breed naturally within the tank, producing a significant quantity of eggs ranging from 0.5-2.5 million.

- Eggs are then incubated in 2-tonne capacity tanks at stocking densities ranging from 200-500 eggs per liter, hatching within 18-22 hours and developing into larvae within 60-72 hours.
- Larval rearing takes place in 2-5 tonne capacity tanks at stocking densities of 5- 10 individuals per liter, with mild aeration. The green water culture technique is employed during this stage.
- Newly hatched larvae are initially fed enriched rotifers, transitioning to inert feed starting from the 15th day and entirely to artificial feed by the 20th day post-hatching. Size grading is conducted every third day to mitigate cannibalism. Larval metamorphosis typically spans 18-21 days.
- The breeding technology is primarily concentrated in the states of Tamil Nadu, Andhra Pradesh, and Kerala.

Silver Pompano *(Trachionotus blochii)*

- Silver pompano, a commercially valuable and high-priced marine finfish, boasts wide distribution along the East and West Coasts of India, as well as in the Atlantic and Pacific Oceans, making it a prime candidate for cage aquaculture.

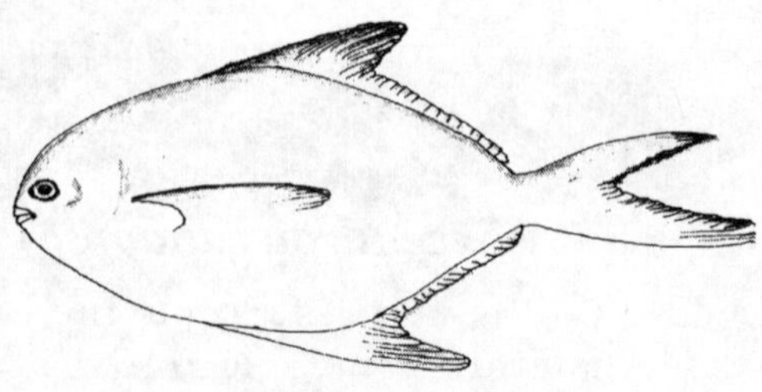

Silver Pompano
(Trachionotus blochii)

- Fish weighing approximately 1.0-1.5 kg are sourced either from the wild or selected from grow-out cages and then stocked in indoor flow-through or re- circulation systems for the purpose of raising gravid brooders.
- Each group of brooders, consisting of one female and two males, is transferred to a 10-tonne capacity FRP tank, maintaining a photoperiod of 15 hours of light at 2000 Lux and a temperature ranging from 27-32°C to facilitate final oocyte maturation.
- Females with intra-ovarian egg diameters exceeding 450 µm, along with males exhibiting milt oozing, are induced to spawn using HCG at 350 IU/kg body weight for both sexes. Alternatively, GnRH can be utilized at a dosage of 150 µg/kg body weight.
- Spawning typically occurs within 36-40 hours post-induction, with hatching taking place between 18 to 22 hours thereafter. A single female fish can lay anywhere between 50,000 to 250,000 eggs, with fertilization rates ranging from 75-90%.

- Floating fertilized eggs are collected using a 500 µm mesh-size net and then stocked in a 2-tonne FRP tank for hatching, with freshly spawned eggs
- measuring 900-1000 µm.
- For larviculture, tanks with a capacity of 5 tonnes are recommended, with a maximum stocking density of 10-15 individuals per liter. The green water technique is employed during this stage.
- Live feeds such as Brachionus and Artemia nauplii are utilized, with the larvae undergoing a color change from black to silvery-white within 18-21 days.
- The breeding technology is predominantly concentrated in the states of Tamil Nadu, Andhra Pradesh, and Kerala.

Indian Pompano *(Trachionotus mookalee)*

Indian pompano, a significant marine food fish found in shallow coastal waters of the Indo West Pacific region, is highly suitable for mariculture.

Indian Pompano
(Trachionotus mookalee)

Breeding occurs predominantly during February-April, with adult males and females weighing between 2.5-3.0 kg raised in Re-circulating Aquaculture Systems (RAS) at a density of 1 kg/m3, and fed with vitamin-mineral mixed squid and clam meat twice daily.

Three months after stocking, females undergo final oocyte maturation, characterized by vitellogenic oocytes measuring more than 500 µm in size, while males reach the milt oozing stage.

Both females and males are injected with human chorionic gonadotrophin at 350 IU/kg body weight to induce spawning, which typically occurs within 36-38 hours post- injection, resulting in the spawning of 0.6-1.5 lakh eggs per spawner.

The eggs hatch after 20-22 hours at temperatures ranging from 28-30°C and salinity levels of 30-32%. Newly hatched larvae measure 2.1-2.2 mm in total length.

Larvae are stocked in green water at a density of 10 individuals per liter and initially fed a combination of rotifers and copepod nauplii (2 individuals/ml), followed by Artemia nauplii and artificial feed.

Metamorphosis of the larvae commences on the 17th day post-hatch and is completed by the 22nd day post-hatch, by which time they reach a size of 1.6-1.7 cm.

The breeding technology is primarily concentrated in the states of Tamil Nadu and Andhra Pradesh.

Orange Spotted Grouper (*Epinephelus coioides*)

- The orange-spotted grouper, a reef-associated marine fish found along the east and west coasts of India, is highly valued for mariculture.
- Breeding seasons are observed during October-November and March-April.

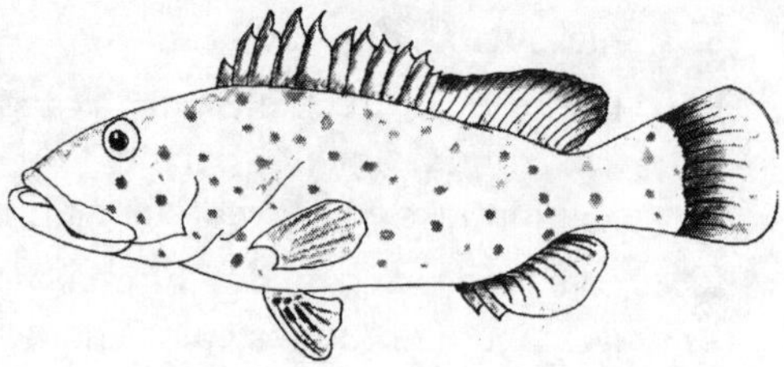

Orange Spotted Grouper (*Epinephelus coioides*)

- Adult groupers weighing 2.0-2.5 kg are raised in concrete tanks at a density of 1 kg/m3, receiving a diet consisting of squid supplemented with a vitamin-
- mineral mixture twice daily in a re-circulating aquaculture system (RAS).
- As protogynous hermaphrodites, males are obtained through hormonal
- implantation using a combination of 17-α methyl testosterone and letrozole at doses of 5 mg and 0.2 mg/kg body weight, respectively.
- Gravid females are identified when approximately 65% of the ova exceed 400 μm in size, while sex-reversed gravid males are identified when they reach the milt oozing stage.
- Natural spawning occurs for 3-15 days each month, with each spawner laying 1-2 lakhs eggs.
- Eggs hatch after 18-22 hours of incubation at 28-30ºC, with average fertilization and hatching rates of 80% and 85%, respectively. The newly hatched larvae measure 1.2-1.6 mm in total length.
- Larvae are stocked in green water at a density of 10 individuals per liter and are fed a combination of rotifers, copepod nauplii, and artificial feed. Metamorphosis into juveniles typically occurs around 33-37 days post-hatch, reaching a size of 1.5-2.0 cm.
- Expansion of this technology is currently underway in Tamil Nadu and Andhra Pradesh.

Giant Tiger Shrimp (*Penaeus monodon*)

- The Giant Tiger shrimp, the largest penaeid shrimp globally, boasts a widespread distribution across the Indo-Pacific region.
- It is a closed thelycum species, with distinguishable sexes characterized by the presence

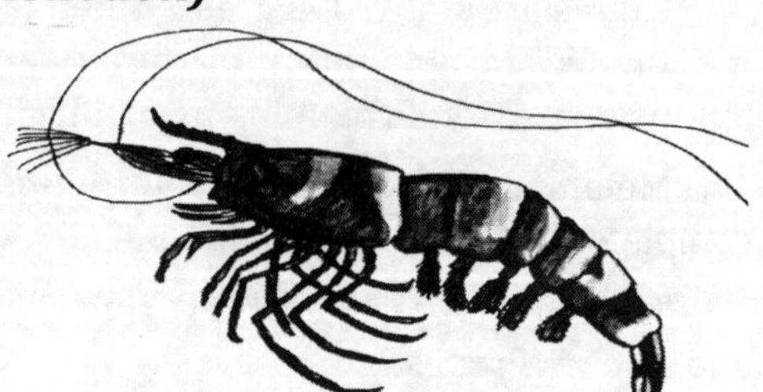

Giant Tiger Shrimp (*Penaeus monodon*)

of petasma in males and thelycum in females. In the wild, first maturity is typically observed at approximately 35 mm (30 g) in males and 46 mm (60 g) in females. However, for hatchery production of seed, larger brooders (>100 g females and >80 g males) are preferred.

- Brooders can also be raised in captivity in maturation tanks, maintained at a density of 4 individuals per square meter with a male to female ratio of 1:2. They are fed live feed such as polychaetes, squid, or maturation feed. Induced maturation is achieved through unilateral eye stalk ablation (ESA).
- Ready-to-spawn females are transferred to spawning tanks. After spawning, the females are removed, and the eggs are incubated for hatching after rinsing several times with clean seawater.
- Eggs hatch within 12-17 hours into nauplii (with a hatching rate exceeding 80%), which are then stocked in larval rearing tanks at a rate of 0.5-1.0 lakh per tonne of clean disinfected water.
- The non-feeding nauplii (6 stages) metamorphose into protozoea stages (3 stages) and then into Mysis (3 stages). Each stage of protozoea and mysis requires 3-5 days to complete its development. They are fed live algae such as Chaetoceros and Skeletonema, while post-larvae are fed Artemia nauplii along with live algae or a microparticulate diet.
- Survival levels until harvest typically range from 30-40%. Post-larvae are reared in the hatchery for an additional 15-20 days before being stocked in ponds.
- This seed production technology is widely adopted in most coastal states.

Indian White Shrimp *(Penaeus indicus)*

- Indian white shrimp, widely distributed in the Indo-Pacific region, typically reaches first maturity at approximately 148 mm and 35 g in females, and 128 mm and 25 g in males. Hatchery operations involve either collecting gravid females from the wild or inducing maturity through unilateral eye-stalk ablation (ESA) and subsequent rearing in captivity until gravid.

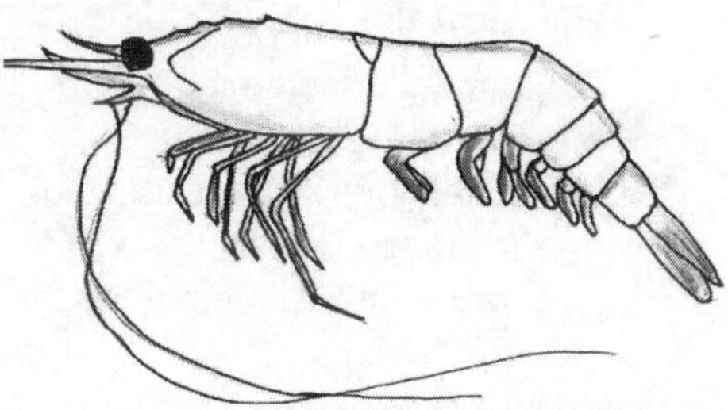

Indian White Shrimp
(Penaeus indicus)

- Brooders undergo thorough disease screening for WSSV, IHHNV, MBV, and EHP before being transferred to the hatchery. After 3-4 days of acclimatization, females undergo ESA for maturation. Broodstocks are reared in FRP/concrete tanks with a capacity of 5 tonnes each, at a density of 6 individuals per square meter, maintaining a male to female ratio of 1:2.

- To produce one million post-larvae, a total tank volume of 80-100 tonnes is required. Ablated brooders are reared in maturation tanks (>5 tonnes) and fed with live feed such as polychaete worms, squids, or molluscan meat, along with pellet feed.
- Females ranging from 35-80 g exhibit fecundity of 1.0-3.5 lakh, with an optimal hatching rate of 85-90%. Nauplii, which are free swimming and non-feeding, undergo six moults (N1 to N6) and are stocked separately in larval rearing tanks at a rate of 1 lakh larvae per tonne.
- Larvae progress through 3 stages of protozoea, 3 stages of mysis, and then to post-larvae, which are reared in different larval rearing sections using microalgae such as diatoms (Chaeoceros, Thalassiosira) during protozoea and mysis stages.
- Live Artemia nauplii (4-5 individuals per post-larvae) and microencapsulated particulate diet are provided during PL1-PL20 stages. By PL15-PL20 stages, they are ready for stocking in ponds. Commercial hatcheries typically achieve survival levels of 70-80% until PL5 stage.
- While seed production technology is adopted to a limited extent, certain hatcheries in coastal states have embraced it.

Banana Shrimp *(Penaeus merguinesis)*

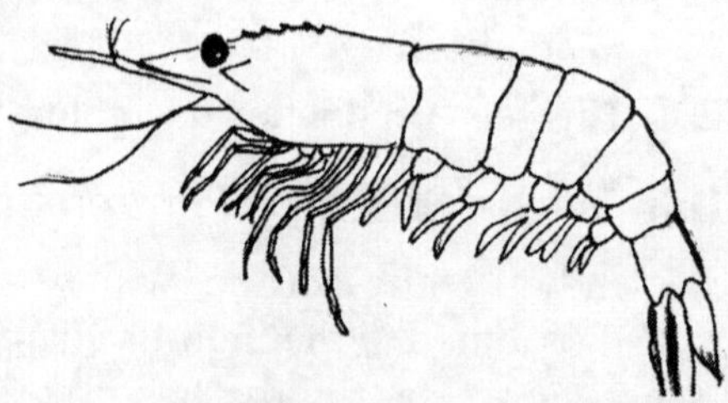

Banana Shrimp
(Penaeus merguinesis)

- Banana shrimp, widely distributed in the Indo-West Pacific region across tropical and subtropical waters, represents a significant fishery resource in South-East Asia and Australia.
- For hatchery seed production, females weighing over 25 g and males over 20 g are utilized, often sourced from the wild. Maturation in adult females can be stimulated through unilateral eyestalk ablation.
- Spawning typically occurs during late night hours, with fertilized eggs hatching within 8-12 hours post-spawning, depending on temperature, with a hatching rate ranging from 80 to 90%.
- Nauplii, the free swimming and non-feeding larvae, are stocked in larval rearing tanks (LRT) at a rate of 50,000 to 100,000 per tonne of water,
- eventually metamorphosing into protozoea through six moults.
- The larvae progress through 3 stages of protozoea, 3 stages of mysis, and then to post-larvae, which are reared in different larval rearing sections,

primarily provided with microalgae such as diatoms at a concentration of 80,000- 130,000 cells/ml during protozoea and mysis stages.

- During the commercial production cycle, an antibiotics-free protocol is followed, with probiotics/bioremediation employed as part of the management strategy.
- Live Artemia nauplii (4-5 individuals per post-larvae) and microencapsulated particulate diet are provided during PL1 to PL12 stages. Post-larvae at PL5 are transferred to PL nursery tanks, and from PL15 onwards, they become ready for stocking in ponds.
- Achieving a larval survival rate of 40-50% from Nauplius to PL5 stage is feasible, with survival rates ranging from 75-85% up to PL15.
- Although the seed production technology exists, its commercial adoption is yet to be realized.

Sand Lobster *(Thenus unimaculatus)*

- The sand lobster, found across the Indian Ocean from West to East, stands out as one of the most promising candidates for lobster aquaculture in India.
- Breeding occurs along the Indian coast from December to March, and successful captive maturation and breeding can be achieved in FRP and concrete tanks ranging from 3 to 20 tonnes.

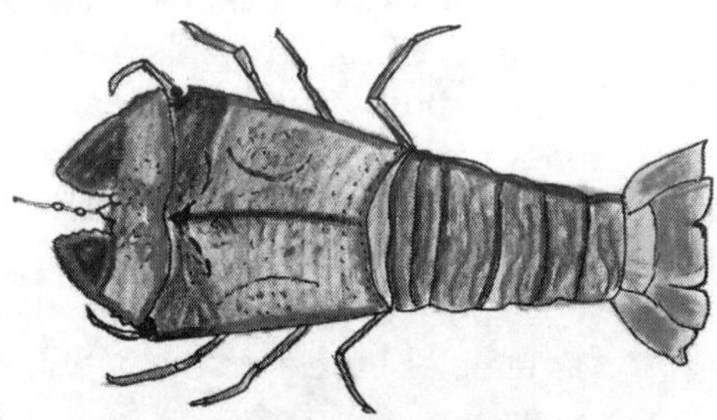

Sand Lobster
(Thenus unimaculatus)

- Sexes are distinct, with mature males identified by a brightly colored genital opening at the base of the 5th walking legs, while females exhibit a general yellowish pigmentation with orange pleopods and uropods.
- Females mature under a 12:12 D:L photoperiod at 26-28ºC. Mating typically takes place in the morning, with fertilized eggs becoming attached to the pleopodal setae for a 5-week incubation period, during which their color
- changes from orange to blackish. A female with a carapace length (CL) of 60- 102 mm can lay 20,000-60,000 eggs.
- Late-stage ovigerous females are transferred to hatching tanks of 200-300 L capacity, equipped with a filter gun and flow-through system to separate hatched phyllosoma stage larvae.
- Phyllosoma larvae begin feeding within 3 hours of hatching and are fed live feeds or particulate diets (such as clam meat) ranging from 400-2000

microns in size, tailored to the molt stages at 6-hour intervals. There are four phyllosomal stages, each taking approximately 4-5 days to molt and grow.

- The fourth phyllosomal stage transitions to the post-larval nisto stage before metamorphosing into seed. This cycle is completed in 26 days under laboratory conditions with water salinity maintained at 35-37 ppt.
- While seed production technology exists, its commercial adoption is yet to be realized.

Indo-pacific Mud Crab *(Scylla serrata)*

Indo-pacific Mud Crab
(Scylla serrata)

- Distributed across the Indo-West Pacific region, spanning from East and South Africa to Southeast and East Asia.
- Mature females are sourced from estuarine/coastal waters or brackishwater ponds, with those above 450 g (carapace width: >12.5 cm) exhibiting a dark orange ovary indicative of fertilized eggs, being transported to hatcheries. Male brooders are unnecessary at this stage, as females have already mated.
- Mature brooders can also be cultivated in hatcheries without inducement. Unilateral eyestalk ablation can expedite maturity, with sand-filled tanks aiding successful spawning.
- Feeding entails offering squid/trash fish at 5-10% of body weight.
- Females typically spawn within a month based on their reproductive condition, with released eggs accumulating under the abdomen as a single egg-mass, known as berried females. Hatching occurs within 10-12 days, with hatched crabs returned to the maturation facility for further spawning.
- Hatched Zoea are reared in larval rearing tanks at a stocking density of 80-100 Zoea/l. This crab undergoes five Zoeal stages (Z1 to Z5), with each stage taking 2-3 days to metamorphose to the subsequent stage. Zoea 5 stage transitions into the megalopa stage, which then takes seven days to convert to Crab Instar 1.
- During the first two Zoeal stages (6-8 days), rotifers are provided at a density of 15-20 individuals/ml. From the third Zoeal stage onwards, newly hatched Artemia nauplii (1- 2 individuals/ml) are fed. Zoea 5 and early megalopa are supplied with 5-7 day old Artemia. Megalopa 3 onwards are cultured in net cages in brackishwater ponds at a stocking density of 50

individuals/m2. These megalopa are fed minced fish and clam meat. Crab instars 2-3 are harvested within 7 days.

- Seed production technology is primarily embraced in Andhra Pradesh and Tamil Nadu.

Chapter-21

Major Candidate Species for Brackishwater Aquaculture

Asian Seabass *(Lates calcrifer)*

The Asian seabass, found across a vast region from the Persian Gulf to Southeast Asia and extending to Australia, undertakes migrations from estuaries to freshwater river areas.

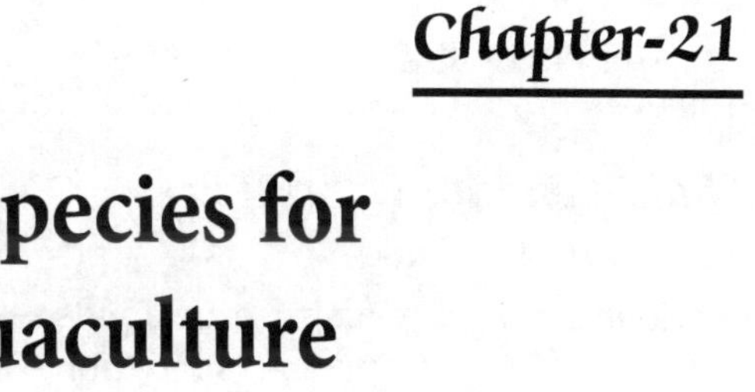

Asian Seabass

This highly predatory fish is typically fed frozen fish like oil sardines and tilapia, constituting 5% of its body weight daily to rear broodstock in captivity.

While Asian seabass exhibits separate sexes, distinguishing between them based on external features is challenging. Males reach maturity before three years of age and subsequently transition to females. They are multiple spawners, releasing approximately 1.0 million eggs per kilogram of body weight in various batches.

Identification of females involves observing a soft, round belly, a pinkish genital papilla, and the emergence of ova upon pressing the belly. Captive breeding involves pairing in a 1:2 female-to-male ratio, selecting females with ova size exceeding 0.450 mm and males showing milt exudation. Spawning is induced by injecting LHRHa hormone, resulting in spontaneous spawning 30-36 hours post-injection.

Fertilized eggs are round and floating, with a diameter of about 0.78 mm, hatching approximately 17 hours later into larvae measuring 1.7 mm.

Newly hatched larvae are stocked in 10-ton indoor FRP tanks at a density of 30 larvae per liter, beginning feeding with rotifers on the second day post-hatch, along with the introduction of green algae. Artemia nauplii and artificial feed are introduced on the 9th and 17th days post-hatch, respectively. By the 30th day post-hatch, seabass fry reach a size of 1.0-1.2 cm with an average survival rate of 40-45%. Due to their cannibalistic nature, frequent grading is necessary to segregate individuals and maintain uniform size.

The breeding and seed production technology for this species is primarily practiced in Tamil Nadu and Andhra Pradesh.

Chanos chanos (Milkfish)

Milkfish is highly favored for brackish water aquaculture in ponds and pens, with widespread distribution across the Indo-Pacific region.

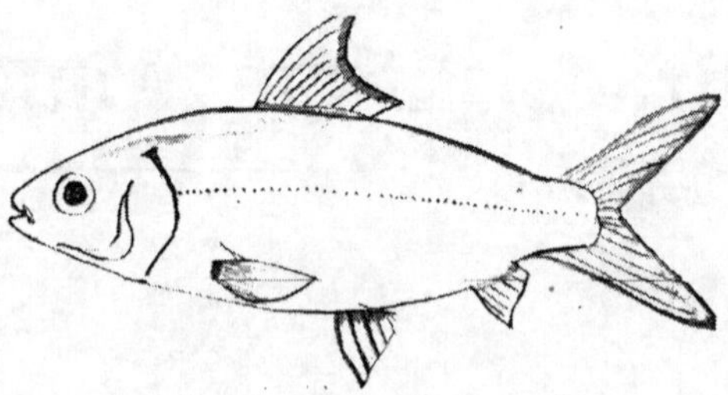

Milkfish

In the wild, they typically spawn twice a year, from January to April (primary season) and October to November. However, in captive conditions, the spawning period can be extended from February to September through hormonal manipulation while maintaining seawater salinity within the range of 30-35%.

Sexual dimorphism is not readily apparent. Fully mature females exhibit ova with a diameter of 650-680 μm during the breeding season, while males release white viscous milt when their belly is pressed.

Broodstock weighing over 3.0 kg are reared in large tanks at a stocking density of 1.0 kg/m3 and fed a floating brood diet to facilitate maturation under captivity. Hormone pellet implantation (LHRH-A + 17α-Methyl Testosterone) is administered at regular intervals to induce spawning in this intermittent spawner.

Mother milkfish can spawn between 0.3-1.0 million eggs per kilogram of body weight. Fertilized eggs are slightly yellowish with a size of 1.10-1.25 mm and hatch within 24- 25 hours at a water temperature of 27.0-29.0°C. Newly hatched larvae measure 3.2- 3.4 mm in total length.

Hatchlings are reared in semi-outdoor tanks at a density of 10-15 individuals per liter. Feeding commences on the second day post-hatch with rotifer Brachionus plicatilis and green algae, followed by the introduction of Artemia nauplii on the 14th day post- hatch. Artificial feed (200-300 μm) is provided from the 20th day post-hatch. By day 30, fry reach a size of 1.5 cm with an average survival rate of 40-45%.

The breeding technology developed is being transferred to Tamil Nadu and other neighboring states.

Scatophagus argus (Butterfish)

- Spotted Scat, also known as 'Butter fish', holds significant importance both as a food source and ornamental fish and is distributed throughout the Indo-Pacific region. This species exhibits euryhaline behavior and is commonly found in coastal and mangrove regions of India.

Butterfish

- Sexes are distinct, with breeding occurring from June to December, characterized by two extended seasons. During the breeding period, mature males (>80 g body weight) and females (>150 g body weight) can be distinguished by the presence of milt exudation and a distended belly, respectively.
- Fully mature females, identified by oocytes exceeding 400 μm, are induced to spawn through hormone administration (HCG and LHRH-a), followed by egg stripping and fertilization by mixing with milt. Fecundity ranges from 1.15 to 1.50 lakh per kilogram of body weight, with pelagic eggs containing a single oil globule and a diameter of 750 μm.
- Hatching occurs within 16-17 hours, resulting in newly hatched larvae measuring between 1.5 and 1.6 mm in size. Larvae are reared in indoor or semi-outdoor larval rearing tanks, where they are introduced to green water and fed with rotifers (2–10 days post-hatch) and Artemia nauplii (11-20 days post-hatch), before transitioning to a microparticulate feeding regime. Within a month of larval rearing in the hatchery, fry typically attain a size of approximately one centimeter.
- Efforts are underway to extend the breeding technology of Spotted Scat to Tamil Nadu.

Terapon jarbua (Crescent Perch)

- Crescent Perch, also known as Target fish, holds significance as both a food and ornamental fish species and is distributed throughout the Indo- Pacific region, ranging from the Red Sea and East Africa to Australia.
- This species is euryhaline, capable of tolerating a wide range of salinities from freshwater to seawater. Spawning occurs in the sea, with juveniles migrating into freshwater for growth.

- Although sexes are separate, differentiation can be challenging. Generally, larger sizes are indicative of females, while smaller sizes are typical of males. Mature males release white milt upon belly pressure, while females are identified through egg cannulation and measurement of ova diameter.

Crescent Perch

- Broodstocks weighing 100-150 g are developed in small net cages (2 m x 1 m x 1 m) placed in ponds or open water bodies at a density of 100 individuals per cage. A 100 g female can release approximately 80,000-100,000 eggs, with hatching observed after 16-18 hours, resulting in hatchlings measuring 2.0 mm in size.
- Larval rearing involves stocking larvae in 10-ton FRP tanks at a density of 10-15 individuals per liter, supplemented with green algae, Chlorella salina, and rotifers (Brachionus plicatilis) as initial feed, transitioning to Artemia salina nauplii later on. Larvae can easily be weaned onto artificial feed at the fry stage. By 45 days, larvae typically reach a fry size of 2.0 cm and weigh between 300-400 mg, with survival rates ranging from 30- 40% achievable through proper feed management, water quality, and health practices.
- Efforts are underway to expand this breeding technology in Tamil Nadu.

Chapter-22

Soil Quality Parameters in Aquaculture

Two fundamental factors crucial for successful aquaculture operations are the upkeep of a healthy aquatic environment and the generation of an adequate supply of fish food organisms, primarily plankton, within ponds. The nutrient levels in both water and soil are paramount in dictating the production of plankton or primary production in fish ponds. Bottom soil regulates the storage and release of nutrients into the water above through a range of chemical and biochemical processes essential for biological production in the environment.

Soil Quality Parameters

Physical Parameters

- Texture of soil

Chemical Parameters

- Total Nitrogen
- Total Phosphorous
- Organic carbon
- Soil PH

Texture of Soil

The relative distribution of mineral components within soil particles (sand, silt, and clay) determines its texture and serves as an indicator of soil's water retention capacity. For aquaculture purposes, pond soilideally should contain at least 35% clay content.

Soil Quality Parameters

Soil

Soil plays an important role in regard to the fertility of fish ponds. Types, characteristics and chemical conditions of soil influences the pond productivity. The physico-chemical properties of pond water are more or less a reflection of the properties of the bottom soil. In this respect the major chemical factors of importance are pH, total nitrogen, total phosphorus, organic carbon, available N_2 and available P.

Hydrogen ion Concentration (pH)

Various factors influence the pH of soil. Low pH levels can significantly impede the release of essential nutrients at the soil-water interface. A pH range of 5.5-6.5 is classified as highly acidic, 6.5-7.5 as moderately acidic, 6.5-7.5 as nearly neutral, and 7.5-8.5 as moderately alkaline, which are considered favorable for fish ponds. pH levels above 8.5 are considered highly alkaline.

Phosphorus

The significance of available phosphorus in soil for enhancing productivity is widely acknowledged. Phosphorus in soil exists in both inorganic and organic forms, with the organic form comprising approximately 35-40% of the total phosphorus content. Soil with available phosphorus (P2O5) levels below 3 mg/100g (30 ppm) is considered indicative of poor productivity, while levels ranging from 3-6 mg/100g (30-60 ppm) are considered average, above 6-12 mg/100g (60-120 ppm) are regarded as high productivity, and levels exceeding 12 mg/100g (120 ppm) are considered excessive.

Nitrogen

Nitrogen in soil is present mostly in organic forms as amino acids, peptides and easily decomposable proteins. The conversion of complex organic forms of nitrogen to simple inorganic forms is carried out by anaerobic microbes. Hence, it is important to know available nitrogen than the total nitrogen in soil. The range of available nitrogen is 50-75mg/100gm of soil relatively more favorable for pond productivity.

Organic carbon

Organic compounds in soil exhibit greater diversity and complexity compared to mineral constituents. Excessive organic content in soil is also undesirable for pond management. Nonetheless, soil with organic carbon levels below 0.5% may be deemed inadequate, while levels between 0.5-1.5% are considered average. Optimal production is typically associated with organic carbon content ranging from 1.5-2.5%.

Physical, Chemical and Biological Factors affecting Productivity of a Fish Pond

Soil Texture: The soil texture, determined by the proportionate composition of sand, silt, and clay particles, serves as an indicator of the soil's water holding capacity. For aquaculture purposes, pond soil ideally contains at least 35% clay.

Water Characteristics

Key Physical Properties of Water Include Depth, Temperature, and Turbidity.

Depth: The depth of water in a pond is crucial as it affects light penetration, which significantly influences pond productivity. In temperate regions, water layers below 3 to 4 meters and in tropical regions below 1.5 to 2 meters have limited biological productivity. Shallow water warms up quickly, providing optimal conditions for aquatic life. Primary production typically occurs from the surface to a depth of 1 meter. However, ponds in tropical regions should not be excessively shallow, as excessively high temperatures can negatively impact production and lead to fish mortality. A water depth of 1.5 to 2 meters is considered ideal for maximizing biological productivity in a pond.

Temperature

Temperature significantly influences the biological functions of fish, including respiration, growth, and reproduction. Fish are highly sensitive to even slight changes in water temperature, with each species having its own optimal temperature range that may vary seasonally. Unlike mammals and birds, fish are ectothermic, meaning their body temperature is regulated by their environment, leading to variations in metabolism. Consequently, water temperature profoundly impacts their life processes.

As water temperature increases, the oxygen consumption of fish also rises, while the oxygen-carrying capacity of water decreases. For example, at 10°C, water can hold approximately 10.92 mg/l of oxygen, whereas at 30°C, it can hold only about 7.53 mg/l. Temperature plays a critical role in the growth and development of fish. While many species exhibit a wide range of temperature tolerance, optimal growth typically occurs within a specific range. Indian major carps, for instance, thrive in temperatures ranging from 18 to 38°C, whereas cold-water species like trout and salmon have lethal temperature limits around 25°C. For carps and Tilapia, lethal limits are approximately 40°C.

Turbidity

Turbidity, resulting from suspended matter in water, can stem from various sources such as clay, silt, organic matter, and plankton. It may be transient due to factors like rainfall, floods, or drainage influx, or persistent due to soil characteristics and continuous wind and wave activity. The measurement of turbidity is determined by the visibility of a Secchi disc, with optimal visibility in fish ponds typically ranging from 40 to 60 centimeters.

Turbidity significantly influences pond productivity. Clearer water allows light to penetrate deeper, promoting plant growth through photosynthesis. Conversely, turbid water reduces light penetration, inhibiting photosynthesis and thereby affecting overall productivity. Additionally, turbid water heats up quickly, trapping nutrients and causing siltation, which can accelerate pond aging. Furthermore, high turbidity levels can suffocate or diminish planktonic organisms. Waters with suspended solids exceeding 400 mg/l are generally not conducive to productivity.

Soil PH

Soil pH plays a crucial role in various soil processes, such as the conversion of soluble phosphates, the effectiveness of different nitrogenous fertilizers, and the adsorption and release of nutrients at the soil-water interface. Optimal bacterial activity in the soil is typically observed at neutral pH levels. For fish ponds, the ideal pH range is between 6.5 to 7.5.

Nitrogen

Nitrogen is essential for stimulating primary production in aquatic environments, with soil serving as the primary source of this element. However, a significant portion of nitrogen remains bound in complex organic compounds. Bacterial decomposition of organic matter in the soil facilitates the release of mineral nitrogen, which can be utilized by aquatic organisms. Adequate nitrogen content, typically in the range of 25 to 50 mg per 100 g of soil, is considered conducive to achieving average production levels.

Phosphorus

Phosphorus is an essential element vital for various cellular functions such as metabolism, respiration, cell division, and growth, as well as protein synthesis and tissue incorporation. Its low availability often limits biological production. Soil typically has a naturally low phosphorus status compared to other major elements. Additionally, phosphate ions have a reactive nature, leading to their conversion into insoluble phosphates, further reducing their availability. As a result, the focus is on assessing the available soil phosphorus rather than the total phosphorus content. An available soil phosphorus level within the range of 3 to 6 mg per 100 g of soil is considered optimal for achieving average soil productivity.

Organic Carbon

Bacterial activity within the soil relies on the carbon content, utilizing it as an energy source. In fish ponds, the natural processes of sedimentation and organic matter decomposition occur, leading to the release of various nutrients from complex organic forms into simpler inorganic compounds. The productivity of both newly constructed ponds and older ponds is influenced by the soil's organic carbon content, with low and high levels being associated with unproductive outcomes, respectively. An organic carbon content ranging from 0.5 to 1.5% is deemed ideal for achieving average fish production levels.

Water

Important chemical properties of water include dissolved oxygen, pH, carbon dioxide (CO_2), ammonia (NH_3), alkalinity, and dissolved organic matter.

Dissolved Oxygen (DO)

The primary deficiency in water often arises from the lack of dissolved oxygen, typically resulting from the decay of organic matter.

Despite oxygen being a major component of air (comprising 20.95%), its solubility in water is limited. This solubility inversely correlates with temperature. Diurnal variations in DO levels are well-documented, with increases during daylight hours due to active photosynthesis, peaking in the afternoon, and gradual decreases from dusk to dawn.

Depletion of DO

The primary cause of DO depletion is the presence of organic matter and its subsequent decomposition. The application of organic manure in high doses should be managed carefully to prevent depletion. Another significant cause is the die-off of algal blooms.

Prolonged exposure to sub-lethal concentrations of oxygen can harm fish, often unnoticed. Sub-lethal concentrations can reduce feed consumption, growth, feed conversion rates, and disease resistance under culture conditions.

DO Requirements of Different Species

The rate of respiration (O_2 consumption) varies depending on species, size, activity level, temperature, and nutritional status. Younger fish, being more active, tend to consume more oxygen. Fish also tend to consume more oxygen when the concentration in the water is high.

A minimum concentration of 5 mg/l is sufficient for warm water fishes, while cold water species require at least 9 mg/l. (Fish can withstand short periods of DO levels falling by 2-3 mg/l below the minimum without fatalities). Experimental evidence has shown that fluctuations in DO, both below and above the optimum range, have adverse effects on growth, appetite, and feed conversion.

Species with accessory respiratory organs, such as Clarias, Heteropneustus, Channa, and Anabas, can survive in poorly oxygenated waters. Tilapia species, for instance, can thrive even at DO concentrations as low as 1 mg/l.

Methods for Increasing DO

Aerators: Increase the air-water interface, creating smaller bubbles with more surface area.

- ❑ Pumping Fresh Water
- ❑ Splashing the Water Surface

pH - Puissance Dihydrogen (Concentration of Hydrogen)

pH, or the concentration of hydrogen ions, serves as a crucial indicator of water conditions in fish ponds. It is defined as the negative logarithm of hydrogen ion concentration, with substances dissolved in water giving it an acidic, neutral, or alkaline reaction.

The pH scale ranges from 0 to 14, with a pH of 7 indicating a neutral reaction. A pH above 7 indicates alkalinity, while below 7 denotes acidity. Alkaline or neutral water tends to be more conducive to productivity than acidic water. The most suitable pH range for fish culture is typically between 6.5 and 9.0. Prolonged exposure to waters more acidic than pH 6.5 or more alkaline than pH 9.5 can lead to diminished reproduction and growth rates. In waters more alkaline than pH 9.5, carbon dioxide becomes less available. Acidic waters can also indirectly impact fish by adversely affecting their food organisms.

Carbon Dioxide

Carbon dioxide (CO_2) is soluble in water, although it constitutes only a minor fraction of the atmosphere (0.03% of air). The primary sources of CO_2 in water include the decomposition of organic matter and the respiration of aquatic organisms such as animals and plants. During photosynthesis, plants utilize CO2 and release oxygen, with CO_2 typically accumulating in water during the night.

CO_2 exists in water in three closely related forms: free CO_2, bicarbonate ion (HCO_3^-), and carbonate ion ($CO_3^{\wedge 2-}$). The concentration of free CO_2 generally remains below 20 mg/l but can reach up to 50 mg/l in organically polluted water. Elevated levels of free CO_2 can interfere with fish respiration, potentially leading to mortality, although fish can tolerate higher CO_2 concentrations if the dissolved oxygen (DO) levels are sufficient.

CO^2 is not highly toxic to fish; however, concentrations between 50-100 mg/l may cause respiratory stress, while levels between 100-200 mg/l can be fatal. Most species can survive for several days with CO_2 concentrations of up to 60 mg/l, provided that dissolved oxygen levels are abundant.

Organic Matter

Organic matter in water exists in various forms, including living plankton, suspended particles of decaying organic matter (detritus), and dissolved organic matter. BOD (Biochemical Oxygen Demand) is often cited concerning organic matter in water.

BOD (Biochemical Oxygen demand)

Biochemical Oxygen Demand (BOD) refers to the oxygen consumed by unstable organic matter during its stabilization in water, facilitated by aerobic bacteria.

In aquaculture waters, the recommended BOD level is 20 ppm, determined by incubating the water for 5 days and calculating the difference between the initial and final dissolved oxygen (DO) levels to obtain a BOD value of 20 ppm.

Hydrogen Sulphide

Fresh water fish ponds should be free from hydrogen sulphide. H2S is produced by chemical reduction of organic matter that accumulates and forms a thick layer of organic deposit at thebottom.

Unionized hydrogen sulphide is toxic to fish, but the ions resulting from its dissociation are not very toxic 0.01 to 0.5mg/l – lethal to fish and any detectable concentration of hydrogen sulphide in water creates stress to fish

Ammonia (NH_3)

The primary source of NH_3 in water is the decomposition of organic manure, feed, and fish excretions. Ammonia exists in two forms: unionized ammonia (NH_3) and ionized ammonia (NH_{4+}). Ammonia concentration refers to the combined levels of unionized and ionized ammonia (NH_{3+} NH_{4+}), with only the unionized form being toxic to fish.

- **0.02 to 0.05 mg/l:** Safe concentration for many tropical fish species
- **0.05 to 0.4 mg/l:** Sub-lethal effects depending on the species
- **0.4 to 2.5 mg/l:** Lethal to many fish species

To maintain a safe NH_3 level, several measures can be implemented:

- Aeration
- Promoting a healthy phytoplankton population to remove NH3 from the water
- Using biological filters to convert NH3 to NO2
- Avoiding excess feeding and opting for high-protein feed
- Avoiding excessive liming practices

Total Alkalinity

Total alkalinity refers to the collective concentration of bases in water, typically expressed in milligrams per liter (mg/l) of equivalent calcium carbonate. Although minor amounts of carbonates of magnesium, sodium, and potassium may have slight effects on the alkaline reserve, the measurement is often expressed in terms of calcium content for practical purposes.

Total alkalinity serves as an indicator of pond productivity. Productive waters typically exhibit alkalinity values up to 100 ppm or 100 mg/l.

- **Less than 20 mg/l:** Stressful for fish
- **20-300 mg/l:** Ideal for fish
- **Over 300 mg/l:** Stressful for fish

Biological Factors

Biological factors influencing fish production are closely tied to the surrounding environment's ability to provide essential food for cultured species. These factors are particularly relevant in rearing operations where no supplementary food is provided, and energy requirements are met through phytoplanktonic primary production. Photosynthesis, driven by light, transforms mineral salts into carbohydrates, providing the necessary energy for plant development. Phytoplankton biomass fluctuates seasonally, typically peaking in spring and summer. Its density is highest in the surface layers of water (0-10m) and decreases with depth. The appearance of different populations is partly linked to water temperature, turbidity, and nutrient depletion.

Diseases in fishes and prawns can be caused by bacteria, viruses, fungi, protozoa, and crustacean parasites. These parasites can enter the pond through water, fish or prawn seed, and nets from other infected ponds.

Excessive growth of aquatic weeds in fish ponds is detrimental to aquaculture systems as these weeds compete with desirable organisms for nutrients.

Estimation of productivity involves estimating primary production, which refers to the rate of organic matter synthesis from inorganic constituents of water by plants (phytoplankton) in the presence of sunlight. Organic production by plants is the initial step in harnessing energy from non-living natural resources, hence termed primary productivity. Methods used for estimation include the Dark and Light bottle method and C techniques.

In the Light and Dark bottle method (Garden and Gran, 1930), the amount of oxygen liberated by phytoplankton during photosynthesis is considered a measure of primary production. Water samples are collected in three BOD bottles: light, dark, and control, at depths where productivity is to be measured. Zooplankton are removed by filtering through a plankton net (300μm). The control bottle is immediately fixed using Winkler's fixatives. The dark bottle is wrapped with aluminum foil and kept in a black bag to protect it from light. The light and dark bottles are then suspended on a raft and anchored. After an incubation period of 4 to 6 hours between dawn to midday or midday to sunset at the respective depths, the bottles are removed and fixed with Winkler's fixatives. The oxygen content in the sample is determined using Winkler's method (or another method of dissolved oxygen estimation). Primary production is calculated using the following formulae:

Net oxygen production = LB-IB

Gross production of oxygen = LB-DB

Where

- LB: Oxygen level in the light bottle
- IB: Initial oxygen level
- DB: Final oxygen level

- 't': Time of incubation
- Gross primary productivity (LB-DB)x1000x0.375 =............................= mgc/ m^3/hr 1.25xt.
- Net primary productivity (LB-IB)x1000x0.375 = ... = mgc/ m^3/hr 1.25xt.

Chapter-23

Water Quality Parameters in Aquaculture

The productivity of a pond relies significantly on key physical parameters such as depth, temperature, turbidity, and light, which collectively shape the water's quality and conditions.

Physical Parameter

Depth: The depth of a pond significantly influences both the physical and chemical characteristics of the water. It dictates factors such as temperature, water circulation patterns, and the level of photosynthetic activity. Shallow ponds allow sunlight to penetrate to the bottom, warming the water and promoting productivity. However, in tropical summers, ponds shallower than 1 meter can overheat, posing risks to the survival of aquatic organisms like fish. Typically, a depth of around 2 meters is deemed optimal for maximizing the biological productivity of a pond.

Temperature : Water temperature is primarily influenced by climate, sunlight, and depth, with the intensity and seasonal fluctuations of temperature significantly impacting productivity. In fish ponds, temperatures typically peak in the afternoon following diurnal fluctuations, with lower temperatures observed during the early morning hours. Tropical water bodies often exhibit higher productivity compared to temperate zones due to increased heat budget. Temperature also plays a crucial role in fish breeding processes, both in natural and artificial environments, and influences chemical changes in soil and water. Decreases in dissolved oxygen (DO2) levels are directly correlated with temperature increases. Fish species demonstrate varying levels of tolerance to temperature, with Indian major carps known for their wide temperature tolerance, termed as eurythermal.

Turbidity: Water turbidity can stem from suspended inorganic particles such as silt, clay, and planktonic organisms. The degree of turbidity fluctuates depending on the basin's characteristics and the sediment carried by inflowing water. Ponds with clay bottoms tend to exhibit high turbidity, hindering light penetration and thereby reducing photosynthetic activity, which acts as a limiting factor for productivity.

Light: Light constitutes another significant physical factor in fish pond ecosystems. The abundance of light energy directly impacts productivity levels. The penetration of light is assessed through turbidity measurements, which reflect the combined influence of various factors including suspended clay and silt, as well as the dispersion of planktonic organisms.

Chemical Parameters

pH: The pH of water is determined by the logarithm of the reciprocal of hydrogen ion concentration, expressed as pH=Log 1(H)+. A neutral pH is 7, with values below indicating acidity and those above indicating alkalinity. Pond water pH undergoes diurnal fluctuations, being alkaline in mid-afternoon and acidic just before daybreak. Optimal fish crop yields are typically associated with water pH levels between 7.0 and 8.0, slightly on the alkaline side. pH levels outside the range of 4.8 to 10.8 are considered harmful.

Alkalinity: The alkalinity or acid-neutralizing capacity of natural freshwater ponds primarily arises from carbonate (CO_3), bicarbonate (HCO_3), and hydroxides of calcium, magnesium, sodium, potassium, ammonium, and iron, with calcium being the predominant constituent. Bicarbonate and carbonate collectively constitute the major portion of pond water alkalinity, expressed as total alkalinity. Typically, calcareous waters with alkalinity levels exceeding 50 ppm demonstrate the highest productivity. Waters with alkalinity below 10 ppm seldom yield significant crops, while those within the 10-50 ppm range may still produce satisfactory results.

Dissolved oxygen: Among the chemical constituents in natural water, oxygen (O_2) holds paramount importance, serving as a key regulator of metabolic processes for both plant and animal communities while also serving as a crucial indicator of water quality. Pond water primarily receives oxygen through two main mechanisms: (1) interaction with atmospheric air on the water's surface, and (2) via photosynthesis. Fluctuations in dissolved oxygen levels occur due to processes such as photosynthesis, respiration, and the slow rate of diffusion. These fluctuations typically peak in the afternoon and decline during the night, reaching their lowest point before dawn. Oxygen concentrations below 3.0 ppm can lead to asphyxia in aquatic organisms due to low O_2 levels, while a minimum of 5.5 ppm of dissolved oxygen is necessary to maintain favorable conditions for a diverse warm water fish population. In ponds with low dissolved oxygen levels, fish may gather near the surface for respiration. For ponds aiming for average or good production, maintaining dissolved oxygen concentrations above 5.5 ppm is recommended.

Hardness: Hardness is the collective measure of soluble calcium and magnesium salts present in the water medium. Typically, in most natural water sources, bicarbonate anions (HCO_3) are commonly associated with cations such as calcium, magnesium, sodium, and potassium. Temporary hardness is primarily caused by bicarbonates of calcium and magnesium, while permanent hardness arises from soluble calcium and magnesium carbonates, as well as salts of inorganic acids such as calcium sulfate ($CaSO_4$). Pond water with a hardness level of 15 ppm or higher is deemed satisfactory for fish growth and does not necessitate lime addition. However, water with hardness below 1.1 ppm requires liming to enhance fish production. Hardness levels below 5 ppm can result in slow growth, distress, and eventual death of fish.

Dissolved Nitrogen and its compounds: The significance of dissolved nutrients, particularly nitrogen, is widely acknowledged, as it plays a crucial role in influencing the growth of phytoplankton in aquatic environments. As a fundamental component of proteins, nitrogen holds a pivotal position within aquatic ecosystems. Ponds with dissolved nitrogen levels below 0.1 ppm typically do not indicate favorable conditions for productivity. In the range of 0.1-0.2 ppm, average production can be expected, while levels above 0.2 ppm are considered favorable. However, the optimal range for nitrogen concentration may fall between 0.3 and 1.3 ppm.

Phosphorus fertility levels below 0.02 ppm are indicative of low productivity, while concentrations ranging from 0.02 to 0.05 ppm typically lead to fairly productive conditions. In the range of 0.05 to 0.10 ppm, productivity is considered good, but levels exceeding 0.20 ppm are deemed excessive. Additionally, apart from absolute concentrations, the ratio between nitrogen and phosphorus concentrations can significantly influence aquatic productivity. The utilization of nitrogen and phosphorus for plankton growth occurs ideally within a ratio of 3:1 to 6:1.

Chapter-24

Feeds and Feeding Management

Nutritional Requirement of Fishes

The nutritional requirements of fish are akin to those of terrestrial animals, necessitating protein, carbohydrates, fats, minerals, vitamins, growth factors, and energy sources for growth, reproduction, and other physiological functions. Deficiencies in any of these essential nutrients can lead to disease or mortality. These nutrients can be obtained from either artificial or prepared diets or natural sources within the aquatic environment.

All animals utilize chemical compounds for energy provision and tissue development. Energy is derived directly from consuming plant material or by ingesting other organisms. These organic substances primarily fall into three major groups-proteins, carbohydrates, and fats-which serve as energy stores.

Finfish necessitate approximately 40 essential nutrients to achieve optimal growth and overall well-being. Among these, proteins, lipids/fats, and carbohydrates are categorized as "macronutrients," while vitamins and minerals collectively constitute the "micronutrients" crucial for promoting healthy fish growth.

Feed ingredients

A wide variety of ingredients are available for use in fish and crustacean feeds

- Grasses
- Pulses and Legumes
- Miscellaneous fodder plants crop
- Fruits and vegetables

- Root crops
- Cereals
- Oil-bearing seeds and oil cakes
- Animal products
- Miscellaneous feedstuff
- Additives

Feed Materials

Feed formation is followed by manufacture; their technology will differ, at least in detail, depending on the type of feed to be manufactured.

Vitamins

Vitamins are vital in trace amounts for fish growth and health, serving key roles in nutrient metabolism. Many water-soluble vitamins act as co-enzymes, supporting various metabolic processes. In intensive fish farming, natural food sources may become limited due to high stocking densities, hence vitamins are typically supplemented as premixes in formulated feeds. Fish obtain some minerals directly through gills and skin, while others are acquired from ingested detritus or natural food.

There are 11 water-soluble vitamins required by fish, including thiamine, riboflavin, pyridoxine, niacin, pantothenic acid, inositol, folic acid, choline, biotin, ascorbic acid, and B12, along with four fat-soluble vitamins: A, D, E, and K. Excess water-soluble vitamins are excreted, whereas excess fat-soluble vitamins in the diet can lead to abnormal growth and liver diseases.

Minerals

More than 20 minerals are recognized for their essential functions in fish physiology. These include calcium, phosphorus, magnesium, sodium, potassium, chlorine, sulfur, and various trace elements such as cobalt, copper, iodine, iron, manganese, selenium, zinc, aluminum, chromium, and vanadium. Calcium and phosphorus are particularly crucial for skeletal development, with calcium also playing roles in blood clotting, muscle function, and nerve impulse transmission, while phosphorus is involved in energy metabolism, cellular membrane permeability, genetic coding, and growth regulation.

Fats

Fats serve as the primary form of energy storage in plants and animals. They contain more energy per unit than any other biological product. Excluding fats typically enhances the palatability of a feed.

Carbohydrates

Carbohydrates represent the most economical and abundant energy source for animals. Plant materials predominantly consist of carbohydrates, ranging from easily digestible sugars to complex cellulose molecules that animals cannot digest. Through symbiotic relationships with bacteria, animals can utilize large amounts of cellular carbohydrates. The value of carbohydrates in fish food is subject to controversy.

Protein

Protein constitutes the structural foundation of the fish body, necessitating a generous dietary supply for optimal growth. However, due to its higher cost compared to carbohydrates or fats, the protein content in the diet is limited to what is necessary for growth and tissue repair, with efforts made to source it from more cost-effective alternatives.

Protein level in Fish Diets

Fish require a higher proportion of protein in their diet compared to non-aquatic animals. For example, the optimal protein content in practical diets for harvested fish typically ranges from 30 to 36%.

Protein Quality

Protein quality is mainly determined by the amino acid composition. Amino acids, the building blocks of proteins, consist of 20-25 types. Fish can synthesize 10 of these amino acids through interconversions or from other intermediary molecules. However, the remaining 10, known as essential amino acids, cannot be synthesized by fish and must be provided in their diet. These essential amino acids include methionine, arginine, threonine, tryptophan, isoleucine, leucine, valine, phenylalanine, histidine, and lysine.

Forms of Feeds

There are two forms of feeds

- Dry feeds
- Non-dry feeds

Dry Feeds

Dry feeds typically consist of dry ingredients or blends of dry and moist ingredients, typically containing 10-12% moisture content, which may vary depending on environmental conditions.

Dry feed is categorized into two groups: (1) Mashes/Meals, which consist of simple mixtures of dry ingredients, and (2) Pellets, where ingredients are compacted using mechanical means to form defined shapes. Depending on the formulation and compacting techniques, pellets may be classified as either floating or non-floating (sinking) pellets.

Non-dryfeeds

Non-dry feeds consist of two types: (1) wet feeds and (2) moist feeds. Wet feeds are typically produced from sources such as trash fish or slaughterhouse wastes, containing moisture levelsranging from 45-70%. Conversely, moist feeds are made from combinations of wet and dry raw materials or from dry ingredients with added water, resulting in moisture contents rangingfrom 18-40%.

Aims of Feed Preparation

Proper nutrition is one of the most important function influenced the ability of cultured organism to attain the potential for growth, repro, and survival. The nutrients requirement varies between sp. and different stage of it life cycle.

Forms of Diets

Diets supplied to aquatic organisms could to vary in form possible diet include.

- Live food generally required for the culture of most aquatic organism at their larval phase.
- Starter feed are generally in the forms of fine consumable or flakes

Basic Steps of Feed Formulation and Processing

(a) Grinding

Grinding reduces particle size and increase the surface area of ingredient, thereby facilitating,mixing, pulleting & digestibility.

(b) Mixing

Ground ingredients are mixed in desired proportion to form hormony blend.

(c) Pelleting

Defined as the compacting of feeds formed by extruding in ingredients or mixture ofingredients feed storage.

Methods of Feed Formulation and Manufacturing

The initial step in diet formulation involves balancing crude protein and energy levels, which can be achieved through trial and error, the square method for either crude protein or energy level, or by solving simultaneous equations. Initially, it's advisable to utilize at least three feedstuffs: one high in protein and energy, one with low or intermediate protein but high energy, and one with low or intermediate levels of both protein and energy. As proficiency in diet formulation grows, any number of feedstuffs can be employed, with space reserved for feed additives such as vitamin or mineral premixes.

The subsequent step is to assess the levels of essential amino acids in the formulation to ensure they meet the animal's requirements. Fish amino acid requirements are typically expressed as a percentage of the diet or as a percentage of dietary protein. Conversion from the percentage of the diet to the percentage of protein is done by dividing the dietary level of each amino acid by the dietary protein level. Although calculating the dietary levels of all essential amino acids may be of interest, it may not be practical at all times. Generally, if arginine, lysine, methionine, and tryptophan levels meet requirements, the other indispensable amino acids are likely to be adequate. However, when unconventional protein sources are used, all ten essential amino acids should be checked.

Should the diet formulation lack any amino acid, a feedstuff containing high levels of that amino acid must be added, substituting another ingredient. Following this adjustment, the protein and energy levels must be re-evaluated to ensure balance.

In practical feed formulation, considerations extend beyond nutrient levels and cost to include pellet quality and acceptability. These factors vary across species and pellet types.

Methods of Feed Formulation

A. Pearson Square Method

In most fish diets, protein is the most expensive component and is typically the first nutrient calculated in diet formulation.

The energy level of the diet is then adjusted to the desired level by adding high-energy supplements, which are less expensive than protein supplements.

The Pearson square method is a straightforward way to determine the probable ratios of mixing different protein ingredients to formulate a feed with the desired dietary protein level.

Procedure for Pearson's Square Method

- ❑ Place the percent of protein needed in the feed ration in the center of the square.
- ❑ Place the percent of crude protein in the basal feed in the upper left corner.
- ❑ Place the percent of crude protein in the protein supplement in the lower left corner.
- ❑ Connect the diagonal corners of the square with lines.
- ❑ Subtract diagonally across the square, preferably subtracting the smaller figure from the larger.
- ❑ Place the result in the opposite corner.
- ❑ Add the figures in the right side corners of the square.
- ❑ Determine the relative percent of inclusion from the total value.

For example, to prepare a diet for carp with 27% protein using rice bran (8.2% protein) and soybean meal (44% protein), construct a square.

Place the protein values of the two feedstuffs in the left corners, with the desired protein level (27%) in the center. Subtract the protein level of the feed from that of the feedstuffs and place the result in the opposite corners from the feedstuffs, ignoring positive or negative signs.

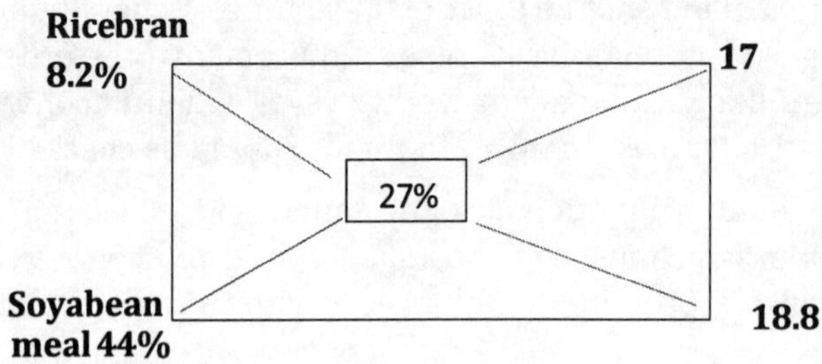

To achieve a 27% crude protein content in carp feed, we must mix 17/35.8 of rice bran and 18.8/35.8 of soybean meal. Next, we need to calculate the proportion of each ingredient by weight in relation to their protein content. This process can be explained as follows:

Rice bran = 17/35.8 ×100 = 47.50	Soyabean meal = 18.8/35.8 ×100 = 52.50

Therefore, to prepare 100 kg of fish feed with 27% protein, you need to mix 47.50 kg of rice bran and 52.50 kg of soybean meal.

B. Linear Programming

In linear programming feed formulation, nutrient requirements are typically fixed to maximize the growth rate, which may not always be the most economically optimal decision. Relaxing nutrient constraints can help reduce feed costs while still maintaining acceptable growth rates.

C. Quadratic Programming

Quadratic programming, on the other hand, considers the growth response within a range of nutrient constraints. This approach requires a thorough understanding of biological response functions derived from actual feeding trials. For instance, by incorporating such data, it may be determined that reducing the inclusion level of arginine by 20% results in only a 5% decrease in the growth of Nile tilapia. Therefore, significant cost savings on arginine supplementation can be achieved with minimal impact on fish growth.

Deteriorative Effects During Storage are Caused by

- Oxidative damage
- Microbial damage
- Insect and or rodent damage/ infestation and
- Other chemical changes during storage

Proper Storage

Effective feed storage practices should safeguard against factors such as high temperatures, humidity, moisture, and infestations by insects and rodents.

Whenever possible, feedstuffs should be stored for the shortest duration feasible. Perishable materials like trash fish ought tobe utilized promptly or kept frozen until needed.

Feeding

Due to its significant cost, careful attention must be paid to providing feed in aquaculture practices. Underfeeding may lead to poor growth, nutritional deficiencies, disease, andmortality, while overfeeding can increase production costs, degrade water quality, induce stress, and impede growth. The quantity of feed administered depends on factors such as total biomass in the pond, expected growth rates, and conversion efficiency. Biomass estimation through periodic sampling guides the determination of feed quantity. Feeding habits dictate feeding schedules, with crustaceans often fed nocturnally or during high tides, while other species are typically fed during daylight hours. Feeding methods include broadcasting or placing feed in designated feeding areas to ensure consumption by cultured organisms. Demand feeders and automatic feeders are employed to enhance feeding efficiency, particularly in fish and crustaceans.

Feed Conversion Ratio (FCR)

The feed conversion ratio, which measures the conversion of feed given into flesh, serves as a key indicator of feed quality and the efficiency of cultured fish in biomass conversion.

$$\text{Food conversion Ratio (FCR)} = \ = \frac{\text{Total Quantity of feed given}}{\text{Total biomass increase}}$$

The best quality feed for a specific cultured species typically results in a low food conversion ratio (FCR).

Chapter-25

Fish Diseases in Aquaculture

Medium for Fish Diseases

Water plays a predominant role in the onset of numerous diseases, as the fundamental physiological processes of fish, including movement, feeding, digestion, assimilation, growth, responses to stimuli, and reproduction, rely heavily on its quality and availability.

Common Symptoms of Diseases

- Anomalous swimming patterns
- Irregular and unhealthy appearance
- Changes in coloration
- Presence of a filmy layer on the skin
- Skin swelling or lesions
- Pallid gills, along with white and red markings
- Excessive mucus production

Sources of Infectious Diseases

- **Primary source:** Affected fish act as the primary reservoirs of pathogens and serve as the main mode of transmission. Pathogens can spread through direct contact or via the release of disease-causing agents into the water.
- **Secondary source:** Water originating from diseased ponds, contaminated sediment, feeds, and equipment.

Natural Resistance of Fish to Infectious Diseases

- The skin surface texture and mucous membrane of fish act as a barrier, preventing infectious microorganisms from entering the body.
- Lysozyme, produced by cells, possesses the ability to destroy bacteria.
- Pathogenic microbes entering the digestive tract are subjected to the action of digestive enzymes, which can effectively neutralize them.
- White blood cells, including lymphoid cells, and reticulendothelial cells in organs like the spleen, liver, and blood vessels, exhibit phagocytic activity, eliminating foreign bodies and pathogens.
- Fish blood contains bactericidin, a compound capable of eliminating various types of pathogenic bacteria.

Types of Diseases

Fish diseases can be categorized into parasitic and non-parasitic types. Parasitic ailments stem from bacteria, fungi, protozoans, worms, leeches, and copepods, while non-parasitic disorders are linked to nutritional deficiencies and abrupt fluctuations in abiotic and biotic factors. Recognizable symptoms in affected fish include deviations from typical behavior, diminished vitality, loss of appetite, and the development of lesions or sores.

Bacterial Disease

Fin and tail rot disease: Caused by *Aeromonas, Pseudomonas,* and *Vibrio bacteria,* Symptoms include white line along the fin margin and brittleness leading to fin ray breakage. Treatment involves 1-minute dip in a 500ppm copper sulfate solution.

- Ulcer disease, triggered by Aeromonas hydrophilla and Pseudomonas bacteria, manifests as open sores or ulcers on the fish's body. Treatment entails a 1-minute dip in a 1:2000 copper sulfate solution for 3-4 days.
- Dropsy, linked to the presence of Pseudomonas punctata bacteria, exhibits symptoms such as fluid accumulation within the body cavity, protruding scales, and exophthalmic conditions. Treatment involves a dip in a 5ppm potassium permanganate solution for 2 minutes.
- Eye disease, attributed to Aeromonas liquifaciens, manifests through symptoms such as vascularization of the cornea, subsequent opacity, and eventual decay of the eyeball. Treatment involves a Chloromycetin bath at 8-10 mg per liter for 1 hour, administered over 2-3 days.

Fungal Diseases

Water mold disease, caused by Saprolegnia parasitica, presents symptoms such as mortality following skin ulceration or shedding, hemorrhagic blindness, and the appearance of white, hair-like tufts in affected areas. Treatment options include a 3-second dip in a 1:10,000 solution of malachite green, or immersion for 5-10 minutes in a 3% common salt solution or potassium permanganate.

Protozoan Disease

Ichthyophthiriasis, commonly known as White-spot disease, is caused by Ichthyophthirius multifilis and is characterized by the presence of small whitish cysts, approximately 1mm in diameter, on the skin, gills, and fins of affected fish. Treatment options include a 5-day bath in 2ppm methylene blue solution, along with hourly dip treatments in a 1:5,000 formalin solution for 7-10 days.

Boil disease, caused by Myxobolus pfeifferi, is characterized by the presence of large boils ranging from the size of a nut to that of a hen's egg on various parts of the fish's body. Treatment involves bathing the affected fish in a 3% common salt solution or a 1:2,500 formalin solution for 10 minutes.

Whirling disease, attributed to Myxobolus cerebralis, is identified by symptoms such as a caudal bend, deformities in the oral region, and darkening of the tail region in affected fish. Treatment involves eradication of all infected fish through the application of quicklime, a pond disinfectant, at a rate of 2 tons per hectare.

Costiasis, caused by Costia necatrix, is characterized by the presence of a bluish coating on the skin and irregular patches of lesions. Treatment involves bathing the affected fish in a 3% common salt solution or a 1:2,500 formalin solution for 10 minutes.

Trematodes

Gyrodactylosis, caused by Gyrodactylus sp., presents symptoms such as fading colors, scale drooping, and skin peeling in affected fish. Treatment involves a dip in a 5% common salt solution or a 1:5,000 formalin solution for 5 minutes.

Dactylogyrosis, caused by Dactylogyrus sp., is characterized by symptoms such as fading colors, drooping scales, and skin peeling in affected fish. Treatment involves a dip in a 5% common salt solution or a 1:5,000 formalin solution for 5 minutes.

Diplostomiasis, commonly known as Black-spot disease, is caused by Diplostomulum sp. and is characterized by the presence of small black nodules, approximately 1-5mm in diameter, in the affected region of the fish. Treatment involves a dip in a 3:1,00,000 picric acid solution for 1 hour, as well as administration of Di-n-butyl tin oxide at a rate of 250mg per kilogram of fish.

Cestodes

Ligulosis, caused by Ligula sp., manifests in fish exhibiting dull, sickly appearances, with portions of the alimentary canal swollen or obstructed by cestode cysts or worms. Treatment involves a dip in a 3:1,00,000 picric acid solution for 1 hour, alongside administration of Di-n-butyl tin oxide at a rate of 250mg per Kg of fish.

Nematodes

Causative organisms such as Philometra sp. and Camallanus sp. lead to symptoms of dullness and sickness in fish, with portions of the alimentary canal swollen or obstructed by roundworms. Treatment includes a dip in a 3:1,00,000 picric acid solution for 1 hour, along with administration of Di-n-butyl tin oxide at a rate of 250mg per Kg of fish.

Acanthocephala

Acanthogyrus sp. is the causative organism behind symptoms such as the appearance of yellowish-white fibro-epithelioma on the lip, skin, and fins of fish. Treatment involves the application of quicklime.

Hirudinea

Hemiclepsis sp. induces abnormal fish movements due to irritation from the parasites feeding on the host's blood. Treatment entails a dip in a glacial acetic acid solution diluted to a concentration of 1:100,000.

Copepods

Causative organisms such as *Argulus* sp., *Ergasilus* sp., *Lernea* sp., *Caligus* sp., *Pseudocycnus* sp., and *Clavellisa* sp. induce symptoms including scale loss, redspot presence, and gill damage in affected fish. Treatment comprises several steps: immersing the fish for half an hour in a 500ppm formalin solution, mechanically removing the parasites using forceps, followed by a 2-3 minute bath in a weak potassium permanganate solution, a 5-minute bath in a 1:1,000 glacial acetic acid solution, and then a 1-hour bath in a 1% common salt solution.Additionally, pond disinfection may be performed by applying lindane at a rateof 8ml per 1,000 liters.

Viral Diseases

Lymphocystis, caused by Lymphocystis spp. (a DNA Iridovirus), manifests symptoms such as lethargy and potential impairment of balance and swimming control, particularly if located along the lateral line. Treatment involves implementing frequent water changes and reducing levels of ammonia and nitrites in the water to alleviate stress, thereby aiding the fish in combatting the infection and naturally reducing tumor size.

Infectious Pancreatic Necrosis (IPN), caused by the IPN virus, is characterized by darker coloration, tail chasing, and spiral swimming behavior in affected fish. Treatment involves ensuring an uncontaminated water supply and providing optimal feed.

Parasitic Disease in Aquaculture

Diplomonads such as spironucleus and kinetoplastids such as cryptobia iubilans. These parasites are commonly found in GI tract of susceptible aquaculture species.

Treatment: Improving environmental conditions and husbandry will decrease morbidity and mortality.

Myxosporean diseases significant in Aquaculture including Whirling disease and Proliferative kidney disease of salmonids and Proliferative gill disease of channel catfish.Whirling disease is caused by myxobolus cerebralis. A presumptive diagnosis of whirling disease is made by detection of spores from skull of infected fish.Diagnosis may be confirmed histologically or serologically.

Proliferative Kidney Disease(PKD): Proliferative Kidney Disease (PKD) stands as one of the most economically significant ailments impacting salmonid industries in North America and Europe, with rainbow trout being particularly vulnerable. This disease stems from Tetracapsuloides bryosalmonae, a myxosporidian featuring four distinct polar capsules. Predominantly observed during summer months when water temperatures exceed 12°C, the parasite predominantly targets yearling and younger fish. Clinical manifestations encompass lethargy, darkening, and fluid accumulation leading to exophthalmos, ascites, and lateral body swelling. Anemia is common among infected fish, evident through gill pallor. Gross examination reveals a notably enlarged, mottled, and gray posterior kidney. Presumptive diagnosis is feasible through the identification of suspect organisms, measuring 10–20 micrometers in diameter, within Giemsa-stained wet mounts of kidney tissue. Confirmation necessitates histologic examination of infected tissue, stained with H&E, and immunohistochemistry. Unfortunately, there is no treatment available, but recovered fish gain resistance to subsequent outbreaks. In nonendemic areas, infected stocks should be depopulated, premises sanitized, and disease-free stock procured for replacement. Prevention remains the most effective measure against PKD.

Miscellaneous Diseases

Gas Bubble Disease

When water becomes oversaturated with nitrogen, usually due to rapid temperature changes, it can lead to gas bubble disease, resulting in significant mortality among fish fry. Affected fish often display symptoms such as swimming at a 45-degree angle with their heads pointing downward, along with the presence of bubbles beneath the skin, on fins, around the eyes, in the stomach and intestine, or within blood capillaries. To address this issue, water agitation is recommended to reduce nitrogen saturation, or alternatively, affected fish should be relocated to other ponds. Additionally, it's worth noting that supersaturated levels of oxygen have also been implicated in causing gas bubble disease in fish.

Vitamin Deficiencies

Symptoms of scoliosis (curved spine) include reduced growth and anorexia (lackor loss of appetite), leading to low fish weight and lethargy.

Chapter-26

Aquaculture Economics

Aquaculture Economics

As aquaculture has grown, the majority of attention has been directed towards biological and technical challenges, largely due to biologists being the primary researchers. Biologists have been concentrating on overcoming production constraints and enhancing traditional systems through modifications and intensification. Various biological issues, such as induced breeding of fish in captivity, disease prevention, understanding nutritional needs of indigenous and exotic species, and selective genetic improvement, are under investigation and continue to be a focus of research efforts.

Despite the increasing focus on biological research, insufficient attention has been directed towards other equally significant areas and their interplay within aquaculture. The viability of aquaculture technology encompasses more than just biology and technology studies; economics plays a crucial role in determining resource allocation efficiency. However, there is often a lack of reliable information on the economics of existing aquaculture systems and the economic feasibility of new technologies. Market evaluation, including demand, marketing infrastructure, and channels, is also vital. Aquaculture Economics is a relatively recent field that aids fish farmers in decision-making and assists in formulating public policies. The potential for aquaculture economics research is extensive, with economists expected to analyze current production and marketing practices, particularly in the private sector, and evaluate emerging husbandry techniques. Economic studies can provide insights into production and marketing aspects of both experimental and existing culture systems, assess aquaculture's role in national economics and international trade, and evaluate development projects and the institutional and cultural context influencing aquaculture development.

Cost Concepts

Cost: Refers to the total amount of funds used in production

Cost of production: Cost refers to the cumulative efforts, sacrifices, and exertion expended in the production of a commodity.

Two major categories of cost namely (1) fixed cost and (ii) variable costs

Fixed Cost

Fixed costs remain constant regardless of the level of output and are incurred even in the absence of production activities.

Fixed cost items – DIRTI-5

Fixed costs are expenses that are not dependent on output levels and therefore do not fluctuate with production quantities. Total Fixed Cost (TFC) comprises various fixed cost components such as depreciation, interest on capital, repairs and maintenance, taxes, and insurance.

Variable Costs

Variable costs correspond to the expenses incurred from utilizing variable inputs, which fluctuate in tandem with production levels. As production increases, variable costs rise accordingly, while they decrease with reduced production. These costs encompass items like seed, feed, manure, fertilizers, fuel, electricity, and labor wages. Variable costs have a more significant impact on production compared to fixed costs.

$$\text{Total costs} = \text{fixed cost} + \text{variable cost}$$

The total cost on a farm, for a specific period, season, or year, represents the monetary value of all inputs utilized. It encompasses both fixed and variable costs, constituting the sum total of expenses incurred. Total cost serves as a crucial metric for computing net revenue.

Total Returns

Total returns are calculated by multiplying the total production by the price per unit.

$$TR = Y \times Py$$

Net returns are determined by subtracting all costs from the total returns.

$$NR = TR - TC$$

Net returns (Variable cost basis)

Equal to the total returns over the total variable costs.

$$NR = TR - TVC$$

Benefit Cost Ratio (BCR)

$$\textbf{Benefit Costratio (BCR)} = \frac{\text{Total returns(TR)}}{\text{Total Cost (TC)}}$$

Benefit cost ratio (BCR) > 1 indicates aquaculture

production is profitable

BCR = 1 (Neither making profit nor less)

BCR < 1 indicates Project is operating a loss

Factors Affecting the Economics of Aquaculture

The producer's profit or net income per unit (Y) is determined by the interplay of production (Q), the cost of production and marketing (C), and the price received (P) as expressed in the equation.

$$Y = QP - C$$

Increasing profits primarily rely on enhancing yield, reducing costs, and raising prices.

The key determinants impacting productivity per unit area include stocking rate, survival rate, and growth rate. The maximum stocking capacity of a fish pond, constrained by limited space and natural food availability, is crucial. Strategies to enhance stocking rate involve fertilization, feeding, polyculture, stock manipulation, and aeration.

Fertilization and Feeding

Utilizing fertilization and feeding typically results in significantly higher production compared to scenarios without these practices. While the total production cost may rise with fertilization and feeding, the cost per unit of fish may decrease, and the additional revenue generated can outweigh the added costs.

Polyculture, the practice of rearing multiple species together, can lead to substantial increases in stocking rates, particularly exemplified by composite fish culture in India. Polyculture enhances production per unit area and yields greater profitability compared to monoculture.

Increase in Farm Prices

In a competitive market price is determined by the supply and demand for fish.

Improvement in the Quality of Fish

Fish is prone to rapid spoilage, resulting in poor quality and waste, ultimately reducing the average price received by farmers. Enhanced fish quality can be achieved through effective preservation methods during transportation and marketing, ensuring that the cost of preservation is outweighed by the additional revenue generated. Graded fish typically fetch a higher price, while marketing live fish tends to increase the price further.

Price fluctuates seasonally : Phased stocking and periodic harvesting offer the potential for price increases.

Co-operative Marketing and Different Markets and Products

Individual farmers often find themselves in a disadvantaged bargaining position, resulting in receiving low prices for their produce. However, by forming farmers' cooperatives and associations, their bargaining power can be strengthened. Furthermore, allocating fish to various markets and selling them in different forms (such as fresh, frozen, salted, and smoked) has the potential to increase prices and maximize revenue.

Reduction in Costs of Production and Marketing

Major production costs in aquaculture are cost of feed, fertilizer, seeds, labor, fuel, electricity,cost of marketing etc.

Cost of Feed and Fertilizers

Feed and fertilizer represent the most significant expenses in aquaculture, often constituting over 50% of total costs. The cost of feed can be mitigated through enhancements in conversion rati os, reductions in unit feed prices, or a combination of both strategies. It is crucial to maintain feed levels at a point where the additional cost aligns with the additional revenue, beyond which it becomes economically unsustainable. Integrated fish farming enhances animal protein production from a given unit area while concurrently reducing feed and/or fertilizer costs.

Cost of Seed

Seed represents a significant portion of total variable costs, especially for species where artificial hatching hasn't been achieved. When supply relies on natural sources, availability fluctuates, leading to price hikes. Breeding in captivity seems to offer a lasting solution. While fingerlings command a higher price than fry, they also boast higher survival rates. Reducing the growth period and stocking larger fingerlings instead of small fry could result in lower seed costs per unit of marketable fish.

Cost of Labor

Labor costs constitute a significant portion of expenses in aquaculture due to the absence of mechanization to replace manual labor. Harvesting, feeding, and maintenance tasks are the primary labor-intensive activities. Thus, efficient management of labor utilization is vital for minimizing product costs.

Cost details for IMC Production

Items	Fixed Cost (€/Acre)
Details of Fixed Cost	
Depreciation cost	9017.19
Interest on capital cost	82598.86
Repairs and maintenance	2040.56
Permanent labour	54021.16
Total fixed cost	**147677.77**
Details of Variable Cost	
Pond Preparation	24874.78
Manuring	10906.53
Seed	49735.45
Feeding	29982.36
Pumping	12405.64
Medicines	1525.57
Pesticides	1497.35
Temporary Labour	9876.54
Marketing	2610.23
Harvesting	2624.34
Total Variable cost	**146038.79**

Economic Details for Carp Farming

Total Fixed Cost (TFC) = ₹ 147677.77

Total Variable cost (TVC) = ₹ 146038.79

Total cost = ₹ 2,93,716.56

Total income = ₹ 3,39,229.45

Net income = ₹ 45,512.89

Benefit cost ratio (BCR) on variable cost = 2.32

Benefit cost ratio (BCR) on total cost = 1.15

Profit margin = 13.41

Fixed cost ratio = 43.53

Variable cost ratio = 43.05

Total cost ratio = 86.58

Chapter-26

Biofloc Technology

Biofloc Technology

Biofloc technology operates as a water-efficient system with minimal exchange, relying on aeration to sustain a dense microbial bacterial floc. Introducing carbohydrates encourages the proliferation of heterotrophic bacteria, which efficiently utilize the ammonia produced by fish, directly contributing to nitrogen uptake. On a dry weight basis, biofloc typically comprises approximately 25-50% crude protein, 2.5% crude lipid, 4% fiber, 7% ash, and offers an energy content of 22 KJ/g.

Biofloc System

Bioflocs represent a complex aggregates of microbes and macroaggregates, encompassing diatoms, macroalgae, protozoa, fecal pellets, exoskeletons, remnants of deceased organisms, organic particles, bacteria, and invertebrates. Typically, these bioflocs found in ponds exhibit a porous and lightweight nature, with diameters ranging from 0.1 mm to a few millimeters. Through controlled conditions within culture tanks or ponds, a symbiotic relationship between heterotrophic bacteria and algae is fostered within these flocs. This microbial biomass thrives on fish excreta, effectively purifying the water by removing undesirable elements. The fundamental operational requirement of a biofloc system is a high stocking density, usually around 3 to 5 kg of fish or 130-150 PL10/m2, alongside robust aeration facilitated by blowers or strategically positioned paddle wheels in ponds. To support growth, pellet grains and molasses are introduced into the culture water. Compared to traditional methods, biofloc systems can significantly boost fish production, with potential enhancements of up to fivefold, while shrimp production typically ranges from 20-25 mt/ha/crop.

Floc Formation

Various binding mechanisms influence the formation, shape, and resilience of flocs. Numerous organisms are enveloped by extracellular polymers comprising polysaccharides, proteins, humic compounds, and similar substances. These viscous polymers serve as adhesive agents, binding cells and other particles together within the floc matrix. Additionally, filamentous organisms play a crucial role in connecting different components of flocs. The composition of biota within the flocs undergoes turnover approximately 2-3 times daily, predominantly consisting of young and metabolically active cells.

Biofloc in Feed and Feeding

Mineral analysis reveals that microbial floc boasts high concentrations of phosphorus along with a diverse array of other minerals. Biofloc encompasses various growth-promoting and probiotic factors, potentially serving as a substitute for vitamins traditionally included in intensive farming systems and fish feed. Feeding tilapia with bioflocs has been observed to double the specific activity of enzymes, suggesting a potential contribution to enhanced growth.

Bioflocs in Disease Management

The heterotrophic microbial biomass is believed to exert a regulatory influence on pathogenic bacteria commonly found in aquaculture systems. Given that intensive farming systems accumulate significant organic matter, primarily from fish biomass, the risk of disease outbreaks is heightened. However, the presence of biofloc can aid in disease outbreak control. Short-chain fatty acids serve as biocontrol agents against pathogenic diseases. Poly-α- hydroxybutyrate (PHB), an intracellular biodegradable polymer produced by a wide array of microorganisms, plays a role in bacterial carbon and energy storage. Upon depolymerization in the fish gut, PHB releases butyric acid, known for its antimicrobial properties. Consequently, PHB may confer a prebiotic advantage in aquaculture. These findings suggest that biofloc represents a novel long-term strategy for disease management, contrasting with conventional approaches like antibiotic, probiotic, and prebiotic applications.

Advantage of BFT

- ❑ Environmentally sustainable aquaculture method.
- ❑ Decreases ecological footprint.
- ❑ Efficient land and water utilization.
- ❑ Minimizes or eliminates the need for water exchange.
- ❑ Boosts productivity, enhancing fish survival rates, growth, and feed conversion efficiency.

- Enhances biosecurity measures.
- Mitigates water pollution and lowers the likelihood of pathogen introduction andtransmission.
- Decreases reliance on protein-rich feeds and reduces feed costs.
- Alleviates strain on wild fish stocks by utilizing more economical food fish and byproductsfor feed formulation.

Species Suitable for Biofloc Technology

Major cultivable fish species in BFT

The selection of species is a fundamental consideration in biofloc system design. Optimal performance is achieved with species capable of deriving nutritional benefits from direct floc consumption.

Biofloc systems are particularly well-suited for species tolerant of high water solids concentrations and poor water quality.

Suitable species for biofloc technology include air-breathing fish such as Singhi (*Heteropneustes fossilis*), Magur (*Clarias batrachus*), Pabda (*Ompok pabda*), Anabas/ Koi (*Anabas testudineus*), Pangasius (*Pangasianodon hypophthalmus*); non-air-breathing fishes like Common Carp (*Cyprinus carpio*), Rohu (*Labeo rohita*), Tilapia (*Oreochromis niloticus*), Milkfish (*Chanos chanos*); and shellfishes like Vannamei (*Litopenaeus vannamei*) and Tiger Shrimp (*Penaeus monodon*).

Chapter-27

Aquamimicry System

Aquamimicry System

Aquamimicry is a concept aimed at replicating natural estuarine conditions by fostering zooplankton blooms, primarily copepods, to provide supplementary nutrition for cultured shrimp, while also leveraging beneficial bacteria to uphold water quality. This involves fermenting a carbon source such as rice or wheat bran with probiotics, typically Bacillus-like organisms, to release essential nutrients. While resembling the Biofloc Technology (BFT), aquamimicry utilizes less carbon input compared to the biofloc system and does not strictly adhere to nitrogen input regulations.

Originating as a response to a shrimp disease outbreak in Thailand during the 1990s, the concept of aquamimicry seeks to replicate natural estuarine conditions by encouraging the growth of zooplankton blooms, predominantly copepods, to provide additional nutrition for cultured shrimp. Moreover, beneficial bacteria play a role in preserving water quality. A carbon source, such as rice or wheat bran, undergoes fermentation by probiotics, such as Bacillus-like organisms, to release nutrients. While similar to the biofloc system, aquamimicry employs reduced carbon input and adopts less strict nitrogen input regulations.

The biofloc and aquamimicry methods have been recognized as effective approaches for shrimp and fish production in the aquaculture industry. Among these methods, aquamimicry stands out as an innovative tool that offers a sustainable solution to challenges faced by the shrimp farming sector. Aquaculture mimicry, or aquamimicry, involves replicating natural estuarine conditions by fostering zooplankton diversity, primarily copepods, to provide additional nutrition to cultured shrimp while also promoting beneficial bacteria in the culture water (Puja *et al.,* 2018).

In this approach, fermented carbon sources like rice bran, combined with probiotics, are utilized to stimulate the growth of phytoplankton and zooplankton populations, mimicking conditions found in natural pond ecosystems. By replicating natural pond conditions, this technology works synergistically to cultivate food organisms similar to those found in natural habitats. Aquamimicry primarily aims to foster the growth of copepods, which are essential zooplankton species serving as supplementary nutrition for cultured species. These plankton organisms contribute to enhancing water quality in fish and shrimp cultures.

Preparation of Pond

The Aquamimicry system consists of three ponds: a sedimentation pond, a biofilter pond, and a grow-out pond. The grow-out pond should be filled with water to a depth of 0.80-1.00 m using a filter bag with a mesh size of 200-300μm. After introducing probiotics (Bacillus) into the pond, dragging a chain across the bottom for a week facilitates soil mixing with the probiotics, minimizing the formation of potentially harmful biofilms for the shrimp. For lined ponds, it is advisable to use heavy ropes instead of chains to prevent damage to the liner. To eradicate small fish or eggs, piscicides can be applied at an ideal concentration of 20 ppm for teased cake or 50-100 ppm for fermented rice bran or wheat bran (without the husk). These additions aid in fostering copepod blooms, and the introduction of additional rice bran can further enhance copepod proliferation within two weeks. Aeration is recommended to ensure proper distribution of nutrients and probiotics in the pond and to mitigate teased levels.

The sedimentation pond should have varying depths, reaching up to 4 m in the center and 2 m at the edges to facilitate sediment accumulation. Bottom-dwelling fish species like catfish or milkfish can be stocked in this pond at low densities, depending on water salinity. Sediments from the grow-out pond stimulate the production of worms and other benthic invertebrates that serve as food for the fish.

Water from the sedimentation pond flows into the biofilter pond, which extends water retention time while functioning as a biofilter. Tilapia can be introduced at low densities in this pond. Subsequently, the water returns to the grow-out pond with minimal nitrogenous waste. Sedimentation ponds should be cleaned every three years. However, a notable drawback of the system is its requirement for a relatively large area, with a pond ratio of 1:1 (treatment to grow-out ponds).

Preparation of Carbon Source

To ensure the effective operation of this technology, it's crucial to provide a carbon source. Rice or wheat bran, minus the husk, should be mixed with water at a ratio of 1:5-10 and combined with probiotics under aeration for 24 hours. The finely powdered bran can be gradually added to the pond, while for the crumbled variety, the upper "milk" or "juice" should be added to the pond, and the bran solids fed to fish in the biofilter pond. Maintaining the pH of the incubation water between 6-7 is important, adjusting it if necessary. Shrimp can be stocked at densities ranging from 30-100 animals per square meter.

For extensive systems, the recommended amount of fermented bran is 1 ppm, while for intensive systems, it's 2-4 ppm, aiming for an ideal turbidity of 30-40 cm. Adjustments should be made accordingly: less bran for higher turbidity and more bran for lower turbidity. As the culture progresses, additional probiotics should be added monthly to support water quality and encourage the formation of biocolloids.

To prevent the formation of biofilms, dragging with chains or ropes on the pond bottom is advised 15 days after stocking the pond with shrimp. While extensive systems may not require further water quality management, excessive sediments should be promptly removed to a sedimentation pond within two hours after each feeding in intensive systems. It's noted that after harvesting, the pond bottom should not have any smell, black soil, or accumulated sediments.

Application of Probiotics in Aquamimicry System

The growth of the aquaculture industry has surged in recent decades, leading to environmental degradation and suboptimal productivity in various crops. To address challenges such as disease resistance, aquatic organism growth, and feed efficiency, probiotics have been increasingly integrated into aquaculture practices. In this system, suitable carbon sources, serving as energy inputs, are combined with probiotics. These probiotics primarily metabolize excess nitrogen, originating from feed and waste, converting it into floc while sustaining copepod populations as a protein-rich feed source (Deepak *et al.*, 2020).

Prebiotics, serving as carbon substrates for probiotic growth, facilitate the establishment of copepod blooms within a week of application. The initial use of probiotics in 1986 aimed to assess their ability to enhance the growth of aquatic organisms. Subsequently, probiotics were employed to combat bacterial infections and enhance water quality in pond systems. Presently, there is compelling evidence suggesting that probiotics can improve reproductive rates, nutrient absorption, and stress tolerance. Commercially available probiotic products derived from various bacterial species such as Carnobacterium sp., Lactobacillus sp., Bacillus sp., Enterococcus sp., and yeast Saccharomyces cerevisiae, among others, are regulated through meticulous management protocols.

Role of Copepods in Aquamimicry Technology

In both intensive and extensive fish culture systems, managing the interactions between fishes and crustaceans, particularly copepods, is crucial for optimizing fish production. Aquamimicry technology, also known as copepod-based production under natural conditions, emphasizes the cultivation of copepods as a primary food source for shrimp species within pond ecosystems, thereby reducing the reliance on intensive mixing, oxygenation, and external feed sources (Romano and Kumar, 2017). Copepods, being crustaceans, are favored for their small size and short life cycle, serving various roles such as food material for marine animals, nutrient recycling, and energy conversion in the food chain (Radhakrishnan *et al.*, 2019; Chakravarty *et al.,* 2018). They are notably rich in LC-PUFA (long-chain polyunsaturated fatty acids) including docosahexaenoic acid, arachidonic acid, and eicosapentaenoic acid, essential for the growth and development of cultured species, offering higher nutritional value compared to rotifers and artemia species (Wilcox *et al.,* 2006; Satoh *et al.,* 2009). Moreover, copepods are known to contain significant amounts of carotenoids, peptides, free amino acids (such as taurine), vitamins, and minerals such as selenium, iodine, copper, and manganese.

Chapter-28

Periphyton Based Aquaculture System

Periphyton Based Aquaculture System

Periphyton a complex assembly of sessile organisms affixed to submerged surfaces like rocks and debris, encompasses algae, invertebrates, detritus, and microorganisms. It serves as a vital dietary source for fish and acts as a conduit for nutrient transfer within aquatic ecosystems. Periphyton communities, contingent upon substrate type, are categorized as 'Epilithon' (rock), 'Epipelon' (mud or silt), 'Epipsammon' (sand), and 'Epiphyton' (submerged macrophytes). Various materials, such as coral reefs, tree branches, aquatic plants, bamboo, and PVC pipes, can facilitate periphyton growth by providing surface area.

Substrate-based aquaculture systems are extensively practiced in Manipur, located in Northeastern India, particularly in Loktak Lake. Throughout the lake, floating islands emerge from dense aquatic vegetation, serving as natural fish gathering points. These circular mats of weeds, known locally as "Phums," contribute significantly to fish production, yielding between 300 to 1000 Kg per Phum area.

Mechanism Involved in Periphyton Based Aquaculture

Two mechanisms have been proposed to explain the increased fish production in periphyton-based systems:

- The presence of additional shelter provided by substrata reduces metabolic costs, enabling a greater allocation of resources towards fish production.
    - ☆ The enhanced primary and benthic production facilitated by artificial substrata establishes a new food web, contributing to fish biomass.

Moreover, filter feeding on phytoplankton is unlikely to meet the micronutrient and energy needs of most herbivorous fishes like rohu. Hence, larger food items such as benthic algae and algal detritus growing on bamboo poles are necessary to supplement the required nutrients for the fish.

Development of Periphyton Assemblage

During the initial phases of periphyton formation, dissolved organic substances are deposited onto clean surfaces, attracting bacteria through hydrophobic reactions. This process is further stimulated by the presence of free-floating organic microparticles in eutrophic water. Bacteria adhere actively using mucilaginous strands, initiating algae growth within days or a week. The succession begins with low-profile diatoms, succeeded by small pennate diatoms, diatoms forming rosettes and mucilage pads, and eventually, species of green algae characterized by upright filaments or long strands.

Periphyton organisms employ diverse attachment mechanisms, such as stalks with adhesive ends, sticky capsules, filamentous cushions, muscular suction pads, adhesive secretions, or direct adherence to the substrate. Attachment to sediment occurs through rooting, rhizoids, or via a muscular foot.

As organism density rises, competition for substrate surface area, carbon dioxide, nutrients, and light intensifies, prompting shifts in the periphyton community composition and its expansion away from the substrate.

Types of Substrate

Utilizing substrates for periphyton growth in aquaculture ponds is recognized to enhance water quality. Common substrates employed include bamboo pipes, plastic sheets, PVC pipes, fibrous scrubbers, and ceramic tiles.

Factor Affecting Periphyton Growth

Factors influencing periphyton growth include the duration of submersion, the type of substrate, light intensity and quality, as well as the depth of the substrates.

Periphyton Community

The Bacillariophyta family, comprising 8 genera, was the most prevalent, trailed by Chlorophyta with 7 genera, and Cyanophyta with 4 genera. Periphyton encompasses phyto-periphyton, zoo- periphyton, and occasionally aquatic insects. Below are some common species found within the periphyton community.

Family	Species
Bacillariophyceae	Fragilaria, Diatom, Navicula, Cyclotella, Pinnularia
Myxophyceae/ Cyanophyceae	Oscillatoria, Rivularia, Nostoc, Anabaena
Chlorophyceae	Spirogyra, Scenedesmus
Eugleophyceae	Euglena
Rotifera	Branchionus angularis
Copepoda	Cyclops, Mesocyclops
Cladocera	Diaphnasoma

Role of Periphyton in Aquaculture

Food Source: Periphyton is essential in freshwater ecosystems, greatly boosting primary productivity and acting as a critical energy source for detritus and grazing food chains. Abundant in nutrients, periphyton serves as a vital food source for diverse aquatic fauna such as fish, crustaceans, snails, chironomids, mayflies, and oligochaetes.

Efficient nutrient utilization: Periphyton has the capacity to utilize nutrients already existing in the water for their growth. Their versatile nature enables them to thrive on various substrates, making it feasible to cultivate them on easily accessible materials to foster their development.

Water Purification: Periphyton serves diverse roles in nutrient management, including the removal of phosphate from the water column via uptake and deposition, as well as the filtration of particulate phosphate. Additionally, these communities aid in nutrient retention by establishing biochemical conditions favorable for phosphate deposition and by trapping particulate material from the water column.

Different Substrates used in Periphyton based Aquaculture system

Fish	Periphyton substrates
Labeo rohita, Catla Catla	Bamboo Poles
Nile Tilapia	Periphyton based fed
Labeo fimbriatus	Sugarcane Bagasse
Litopenaeus vannamei	Artificial Substrate

Conclusion

Periphyton plays a crucial role in phosphate removal from surrounding water by absorbing and depositing phosphate, as well as filtering particulate phosphate from water bodies. This process demands minimal energy inputs and can be achieved using cost-effective materials. Given that filter feeding on small planktonic algae is insufficient to meet the energy needs of herbivorous carp and tilapia, farmers have the potential to enhance productivity, lower operational expenses, and broaden income sources through periphyton-based aquaculture. However, careful management and monitoring of water quality and nutrient inputs are essential to sustain healthy periphyton growth and prevent adverse impacts on the ecosystem. Utilizing periphyton as a food source is sustainable, as it is naturally occurring and doesn't rely on expensive or environmentally harmful inputs.

Chapter-30

Flocponics System

Flocponics revolutionizes aquaculture by blending recirculating aquaculture systems (RAS) with hydroponics, fostering a symbiotic relationship between fish and plants. Waste produced by fish is converted into ammonia, then nitrate by nitrifying bacteria, enriching water transferred to hydroponic setups. This mutualistic setup not only boosts nitrate removal and water quality but also minimizes environmental footprint, offers flexibility, bolsters fish health, and slashes labor and energy expenses compared to conventional aquaponics systems.

Flocponics represents a pioneering fusion of Recirculating Aquaculture Systems (RAS) and aquaponics, merging the water treatment and recirculation methodologies of RAS with the nutrient recycling and plant cultivation components of aquaponics.

Within Flocponics, fish thrive in a meticulously managed RAS environment, where water undergoes treatment and recirculation to ensure ideal conditions for fish well being and development. The waste generated by the fish undergoes a process wherein it is transformed into ammonia, subsequently decomposed by nitrifying bacteria into nitrite and nitrate.

Subsequently, the water rich in nitrate is directed to a hydroponic setup, facilitating soil-less cultivation of plants. These plants utilize the nitrate as a nutrient source, efficiently extracting it from the water. This fosters a mutualistic bond between the fish and the plants, where the fish contribute nutrients to the plants, while the plants play a pivotal role in purifying the water for the fish.

Flocponics offers several advantages over traditional aquaponics systems:

- **Enhanced Nitrate Removal:** Flocponics, employing a RAS setup for fish cultivation, excels in nitrate extraction from water, surpassing conventional aquaponics systems. This capability allows for higher stocking densities and amplified production yields.
- **Elevated Water Quality:** The integration of RAS and hydroponics in Flocponics leads to heightened water quality for both fish and plants. Hydroponic vegetation aids in nutrient regulation, while the RAS system maintains consistent and optimal conditions essential for fish health and development.
- **Mitigated Environmental Footprint:** Flocponics mitigates the environmental impact of aquaculture by proficiently recycling and reusing water, minimizing waste generation, and reducing reliance on external inputs like fertilizers.
- **Enhanced Adaptability:** Flocponics showcases increased adaptability to diverse fish species and plant varieties, rendering it a versatile and flexible option for aquaponic endeavors.

Aquaponics and biofloc-based aquaculture are recognized for their environmentally friendly approaches to food production, emphasizing nutrient recycling and water conservation. Flocponics aligns with these principles, merging the concepts of aquaponics and bioflocs. By integrating these methodologies, flocponics presents an additional avenue to address the challenges of global sustainable food supply. Pinho *et al.* 2021 recently introduced the term 'FLOCponics' to encompass and standardize systems previously referred to as 'BFT+hydroponics,' 'BFT+aquaponics,' or 'BFT+plant production'.

Challenges of Flocponics System

Flocponics represents a multifaceted food production system requiring extensive expertise spanning microbiology, limnology, ecology, aquaculture, engineering, agronomy, and hydroponics. Given its intricate nature and limited research to date, there are notable knowledge gaps regarding flocponics that necessitate further investigation. At this nascent stage of scientific exploration, identifying and addressing the challenges while highlighting the opportunities inherent in flocponics can offer valuable guidance for future research endeavors, ultimately facilitating its efficient advancement. Presently, technical hurdles pose challenges to the sustainability and economic viability of flocponics.

Once the system design enhancements are integrated, the primary challenge in animal production within flocponics lies in ensuring an optimal presence of in situ food/bioflocs within the aquaculture tanks. Upon implementation of the aforementioned improvements, the complete nutritional benefits of flocs can be realized. Several nutritional benefits associated with utilizing Biofloc Technology (BFT) instead of Recirculating Aquaculture Systems (RAS) have been reported, including:

- Decreased feed conversion ratio.
- Substitution of fish meal with alternative protein sources.
- Lowering of dietary protein content, particularly in tilapia and Pacific white shrimp farming.

Both these species are extensively cultivated and play significant roles in global food provision. While Flocponics may face limitations in the production of other high-value species, focusing on well- established products during the development of new technologies can be advantageous.

Application of Flocponics

Flocponics is poised to become a viable alternative for traditional aquaponists or investors seeking to establish integrated agri-aquaculture farms, once technical barriers are addressed. Currently, a diverse range of knowledge is essential to effectively operate a FLOCponics system and optimize its outcomes. Additionally, selecting the appropriate food production system involves considering numerous factors such as market demand, climate conditions, producer expertise, technical proficiency, input costs, and availability. Even with the anticipated benefits of flocponics, a comprehensive analysis of the entire production landscape is necessary to determine the most suitable system for specific circumstances. Unlike traditional aquaponics systems, typically operated on a small-scale for personal or family use, flocponics leans towards medium and large commercial-scale operations due to the complexity of Biofloc Technology (BFT) and the need for substantial infrastructure and investment. Therefore, it's unlikely that flocponics will be commonly employed as a backyard system, underscoring the importance of enhancing and standardizing system designs for practical production settings.

Chapter-31

Aquaponics System

In aquaponics, the effluents from Recirculating Aquaculture Systems (RAS) serve as nutrient-rich resources for plants (Paudel, 2020; Wongkiew *et al.*, 2017). However, these systems can be costly due to the installation requirements for both RAS and hydroponic setups (Palm *et al.*, 2018). Plants that thrive in hydroponic environments typically do well in aquaponic setups. If the nutrient levels in the RAS effluent are insufficient, additional fertilizers can be applied (Eck *et al.*, 2019; Maucieri *et al.*, 2019). Unlike coupled aquaponic systems where water flows through all components, decoupled systems allow for adjusting water flow through fish tanks, filters, and hydroponics (Goddek *et al.*, 2016; Monsees *et al.*, 2017). Aquaponic systems have the potential to minimize nutrient losses from aquaculture. A recent innovation is the flocponic system, which replaces RAS with a biofloc system connected to hydroponics (Pinho *et al.*, 2017). Key challenges for profitability in aquaponic or flocponic systems include accessing high-value markets for vegetables and fish, requiring diverse knowledge, and emphasizing the importance of vegetables for financial success (Bosma *et al.*, 2017). These factors contribute to the slow adoption of these systems in the industry.

Aquaponics Definition

Growing Fish and Plants Together Aquaponics Defined as:

The integration of:

Aquaculture – *Growing fish in a* re-circulating system

Ponos – *The Greek word for growing* plants with or without media

How Aquaponics Works

- ❑ Fish inhabit a tank.
- ❑ Water from the fish tank is transferred to nourish the plants.
- ❑ Bacteria facilitate the conversion of ammonia and nitrite into nitrate.
- ❑ Plants uptake the nutrient-enriched water.
- ❑ Purified water is reintroduced into the fish tank.

In aquaponics, plants are cultivated in the grow bed, while fish are housed in the fish tank. The nutrient-rich water, containing fish waste, is directed to the grow bed, where billions of beneficial bacteria naturally present break down the ammonia into nitrites, and subsequently into nitrates. Plants utilize these nitrates and other nutrients for growth. The roots of the plants act as filters, cleansing the water before it cycles back into the fish tank, ensuring a continuous flow of fresh, oxygenated water for the fish. This cyclical process maintains water quality and fosters symbiosis between fish and plants in the aquaponics system.

Aquaponics Components

- ❑ Fish Tank
- ❑ Place to Grow Plants
- ❑ Water Pump(s)
- ❑ Air Pump
- ❑ Irrigation Tubing
- ❑ Water Heater
- ❑ Filtration (Optional)
- ❑ Grow light
- ❑ Fish and Plants

Safe Materials-Make sure all system components are fish and human safe

- ☆ Polypropylene - labeled PP
- ☆ High Density Polyethylene - labeled HDPE
- ☆ High Impact ABS (Hydroponic Grow Trays)
- ☆ Stainless Steel barrels
- ☆ EPDM or PVC (poly vinyl chloride) pond liner (makeits UV resistant and avoid fire retardant material)
- ☆ Fiberglass tanks and grow beds
- ☆ Rigid white PVC pipe and fittings, black flexible
- ☆ PVC tubing, some ABS
- ☆ DO NOT use Copper – Its toxic to the fish

Advantages

- Aquaponics presents numerous environmental advantages compared to hydroponics, aquaculture, and traditional agriculture.
- Aquaculture often confines fish tightly in densely populated pools, leading to significant volumes of waste-laden water.
- Aquaponics obviates the need for wastewater treatment and disposal by recycling and filtering water within the system.
- Nitrogen fertilizers necessary for plant growth in hydroponics are supplied by fish in aquaponics systems.
- Inefficient irrigation is a primary cause of water wastage in conventional agriculture, with substantial freshwater lost through evaporation or runoff before reaching crop roots.
- Closed aquaponics systems, however, achieve efficient crop growth with minimal water wastage and low water loss.

Disadvantages

The expense has posed a substantial barrier to the adoption of commercial aquaponics ventures, as some farmers find the initial investment and ongoing operational costs, including fish feed and pump electricity, to be prohibitive. Moreover, the systems require significant labor input, necessitating regular maintenance and expertise in both plant and fish care. Research suggests that aquaponics systems may not generate as much profit as traditional agricultural methods. Nonetheless, proponents of aquaponics contend that these assessments reflect the early stages of the technology and assert that profitability is attainable in adequately sized facilities.

Chapter-32

Integrated Multitrophic Aquaculture System

Integrated Multitrophic Aquaculture System

Integrated multi-trophic aquatic systems represent a modern and more sustainable approach to production by combining fed species with extractive species, mimicking natural ecosystems. In this system, nutrients and by-products from fed culture are utilized by extractive species for their growth, promoting nutrient reuse and reducing environmental impacts, fertilizer, and water costs, thus fostering circular food production. Furthermore, integrating systems and species with different trophic functions enhances product variety and local food security. Aquaponics exemplifies this concept by merging aquatic animal and vegetable production in a single loop. While the traditional setup involves freshwater RAS and hydroponics, emerging variations such as decoupled systems, multi-loop designs, algaeponics, maraponics, and flocponics-where biofloc- based aquaculture replaces RAS-are gaining popularity. For instance, flocponics offers an alternative aquaponics system utilizing BFT.

In an Integrated Multi-Trophic Aquaculture (IMTA) operation, careful selection and placement of various components or species are crucial. The system design must be engineered to efficiently recapture waste products. Larger organic particles, such as uneaten feed and feces, settle within the cage system and are consumed by deposit feeders like sea cucumbers and sea urchins. Concurrently, fine suspended particles are filtered out of the water column by filter-feeding animals such as mussels, oysters, and scallops. Seaweeds are strategically positioned a bit further from the site, in the direction of water flow, to effectively remove inorganic dissolved nutrients like nitrogen and phosphorus from the water. In essence, the IMTA system functions by utilizing waste materials from fed aquaculture species, which are then consumed by organic extractive aquaculture species, subsequently releasing gases that are utilized by inorganic extractive aquaculture species. It is essential for IMTA species to be economically viable as aquaculture products and cultured at densities optimized for the uptake and utilization of waste material throughout the production cycle.

IMTA

In recent years, the concept of integrated aquaculture has gained momentum, particularly in Asian regions, as a response to excessive nutrient and organic loads generated by intensive aquaculture. To enhance the effectiveness of an Integrated Multi-Trophic Aquaculture (IMTA) system, the community composition should include not only main species like finfish or shrimp, along with their specific feed, but also both filter and deposit-feeders, facilitating the transfer of products from one species to serve as inputs for another. IMTA entails cultivating species that require feeding alongside extractive species that utilize the organic and inorganic waste generated by aquaculture for their growth. As outlined by Barrington (2009), IMTA represents the strategic integration of fed aquaculture species (e.g., finfish/shrimp) with organic extractive aquaculture species (such as shellfish/herbivorous fish) and inorganic extractive aquaculture species (e.g., seaweed), establishing balanced systems that promote environmental sustainability (through mitigation), economic stability (via product diversification and risk reduction), and social acceptability (enhanced management practices).

Design of IMTA

An effective Integrated Multi-Trophic Aquaculture (IMTA) operation requires meticulous selection, arrangement, and positioning of various components or species to efficiently capture both particulate and dissolved waste materials from fish farms. Careful consideration of species and system design is essential to optimize waste reclamation. Larger organic particles, such as unconsumed feed and feces, settle below the cage system and are consumed by deposit feeders like sea cucumbers and sea urchins. Concurrently, finer suspended particles are filtered out from the water column by filter-feeding organisms such as mussels, oysters, and scallops. Seaweeds are strategically located farther from the site, in the direction of water flow, to target specific inorganic dissolved nutrients like nitrogen and phosphorus. The selected IMTA species should be economically viable as aquaculture products and cultured at densities that maximize the absorption and utilization of waste materials throughout the production cycle. IMTA systems can be land-based or open-water setups, suitable for both marine and freshwater environments, and may involve various species combinations.

Species cultured in Integrated Multi-Trophic Aquaculture (IMTA) encompass various trophic levels, including fed, extractive, and deposit feeders. IMTA systems can be diverse, ranging from land-based to open-water systems, in marine or freshwater environments, and can incorporate numerous species combinations. Examples of IMTA combinations include shellfish/shrimp, fish/seaweed/shellfish, fish/shrimp, and seaweed/shrimp. IMTA is recognized as a more sustainable approach compared to traditional monoculture systems, which rely on a single species. Monoculture practices often have adverse effects on local ecosystems due to their reliance on external food and energy sources, often lacking sufficient mitigation measures.

In IMTA systems, finfish are typically the primary fed component, providing nutrient energy introduced by humans. Fish contribute dissolved and particulate nutrients, as well as compounds that aid in reducing oxidation-reduction potential, benefiting other organisms within the system and the industry's revenue. The nutrient specifics, including quantity and type, are influenced by factors such as species, size, and feed composition. Extractive species in IMTA include organic (e.g., bivalves/shellfish or herbivorous fish) and inorganic (e.g., seaweeds or aquatic plants) extractive species, working together to maintain ecological equilibrium.

In the organic extractive species subsystem of IMTA, filter-feeding bivalves are often cultivated near meshed fish enclosures in an open-water integrated culture arrangement. This setup reduces nutrient loads by filtering and assimilating particulate waste, including fish feed, feces, and phytoplankton growth stimulated by dissolved nutrient waste. Seaweeds are particularly suitable for biofiltration due to their high productivity and economic feasibility for cultivation.

In contrast, deposit feeders are more prominently included in IMTA systems in regions like Japan and parts of Asia, where they are in high demand. Species such as holothurians, or sea cucumbers, can consume a significant portion of settling organic material, effectively removing organic carbon and nitrogen waste originating from fish production. With low mortality rates and rapid growth, sea cucumbers can reach market size in 3 to 4 years, making them viable secondary species in the context of IMTA.

The benefits of IMTA include

- Biological means are employed for effluent mitigation, utilizing bio-filters tailored to match the ecological characteristics of the aquaculture location. This approach not only manages effluents effectively but also enhances profits through diversification. Commercially grown by-products contribute to increased economic gains, leveraging the extractive capabilities of co-cultivated lower trophic level species to acquire supplementary items that outweigh the expenses of establishing and operating an IMTA farm. Surplus nutrients within integrated aquaculture are utilized to support bio-filters, further optimizing resource utilization.

- Additionally, IMTA fosters economic growth by boosting the local economy, creating direct and indirect job opportunities, as well as facilitating processing and distribution of products. Moreover, IMTA serves as a form of inherent risk mitigation, as diversifying products provides financial security and reduces economic vulnerabilities in the face of price fluctuations or crop loss due to disease or adverse weather conditions.
- Furthermore, IMTA contributes to disease management, with certain types of seaweed possessing antibacterial properties that can prevent or reduce fish diseases caused by fish- pathogenic bacteria. Finally, IMTA operations have the potential to enhance profits through premium pricing by participating in eco-labeling initiatives or securing organic certification, thus adding value to IMTA products in the market.

Chapter-33

Organic Aquaculture

The worldwide fishing industry confronts a critical predicament stemming from overfishing, environmental pollution, and the decline of vital habitats like mangrove forests and coral reefs. Moreover, intensive aquaculture methods have resulted in the buildup of detrimental substances such as antibiotics, PCBs, pesticides, and heavy metals in aquatic ecosystems, inflicting considerable harm.

Organic aquaculture presents a sustainable and eco-friendly approach, utilizing careful management of natural resources and ecosystems to yield premium food products. However, a significant challenge in contemporary aquaculture methods involves dependency on wild fish as feed, resulting in the depletion of natural stocks. Furthermore, the farming of fast-growing exotic fish species frequently contributes to the weakening of indigenous species and the transmission of diseases to wild populations.

Organic aquaculture seeks to address these issues by establishing environments that closely resemble natural habitats for cultivating aquatic organisms. Despite consumer demand, numerous fish farmers have been hesitant to embrace organic standards, contending that while current aquaculture practices align with organic principles, they may not meet stringent legal requirements.

Organic aquaculture involves the production of high-quality foods within a stable aquatic ecosystem, achieved through the careful management of natural resources and environment to prevent adverse impacts. Its aim is to safeguard genetic diversity and the abundance of species within their native systems.

Principles of Organic Aquaculture

Key components of organic aquaculture include rigorous monitoring of environmental impact, adoption of natural breeding methods without reliance on hormones and antibiotics, exclusion of inorganic fertilizers, incorporation of natural plant communities into farm management, avoidance of synthetic pesticides and herbicides, sourcing feed and fertilizer from certified organic agriculture and fisheries, adherence to organic sustainability criteria for fishmeal sources, prohibition of GMOs in stocks and feed, implementation of stocking density limits, regulation of energy consumption (e.g., for oxygenation), preference for natural medicines, and processing in approved organic facilities.

History of Organic Aquaculture

Organic aquaculture, stemming from organic farming practices, has undergone significant advancements since its inception in Austria and Germany in the early 1990s. During this period, although the organic food market was burgeoning, organic fish remained relatively scarce. The successful introduction of organic salmon in Germany, the United Kingdom, and France played a pivotal role in driving the global expansion of organic aquaculture. Another significant milestone was the emergence of organic shrimp production, which garnered international attention and spurred European development agencies to extend these initiatives to regions in the global south. Throughout Europe, Asia, and Latin America, various finfish species, such as trout, Pangasius catfish, tilapia, seabass, and seabream, have been cultivated using organic methods. By 2009, there were 240 certified organic aquaculture facilities across 29 countries, predominantly in Europe. In China, 72 operations are certified under national regulations, focusing mainly on carp polyculture, often in conjunction with crabs, shrimps, or indigenous species. Additionally, countries like Bangladesh, India, Thailand, and Vietnam have witnessed an uptick in organic Black Tiger Shrimp production, while Vietnam has also made strides in organic Pangasius catfish farming. Furthermore, China has obtained certification for organic production of turtles and sea cucumbers.

In India, organic aquaculture encompasses the cultivation of micro-algae, yielding a total output of approximately 53,500 tons in 2009, representing about 0.1 percent of global aquaculture production. INDOCERT, a certification body in India, is authorized by the National Programme for Organic Production (NPOP) under the Government of India to conduct inspections and grant certificates for organic production systems. INDOCERT offers certification for the domestic market according to National Standards for Organic Production and for export markets based on the standards set by the United States Department of Agriculture (USDA).

General Standards for Organic Aquaculture Production

Some of the general standards which are to be followed in an organic farming aquaculture system are listed unit wise

Production Unit Setup

Standards: It is imperative for the production unit to effectively manage the surrounding environment to prevent any adverse impacts, such as the escape of cultured organisms or the spread of diseases, while also refraining from the use of synthetic and chemical fertilizers and paints. Prioritizing the surrounding environment is essential for both the placement and the overall management of the organic unit.

Environment / Water Quality

Standards: Aqua farmers must minimize feed wastage and fecal matter in the water system to prevent over-fertilization or other disruptions to the natural environment. It is crucial for aqua farmers to avoid depleting or excessively exploiting available water resources and to uphold the natural water quality.

Breeds and Breeding

Standards : Utilize breeds adapted to local conditions, and ensure breeding is conducted with a large number of pairs to mitigate risks of inbreeding, genetic damage, and loss of genetic diversity. Avoid the use of triploid, genetically modified, and sex-reversed organisms.

Feed and Feeding

Standards: Feed should primarily comprise organic raw materials sourced from wild aquatic stocks. To effectively manage fishery resources, aquatic raw materials from stocks not intended for human consumption and from by- products must be utilized for feed preparation. Feeding practices should ensure that natural feed available within the pond system is also consumed, minimizing wastage of supplemented feed.

Additives

Standards: Avoid the use of additives like growth promoters, hormones, and appetizers in organic farming practices. When incorporating usable additives such as vitamins, minerals, antioxidants, and coloring agents, prioritize those derived from natural sources or as close to their natural form as feasible.

Synthetic and unnatural additives are strictly prohibited in organic farming systems.

Treatments and Animal Health Welfare

Standards: Prioritize the "prevention is better than cure" approach in health management for aquatic organisms to minimize the need for medication. If signs of disease persist, promptly implement appropriate measures. Utilize medications with minimal environmental impact and low risk to human and animal health for treatment.

Record Keeping

Standards: Adhering to strict protocols in both feed and disease management is crucial for maintaining the health of aquatic organisms. Detailed records documenting disease occurrences and treatment procedures must be diligently maintained. Additionally, farm managers are responsible for keeping monthly records of the feed type, producer, and quantity utilized throughout the crop cycle.

The absence of a universally acknowledged standard for organic aquaculture practices presents a barrier for investment in this sector. To attract investment and effectively communicate the benefits of organic aquaculture products to consumers, formally recognized standards are essential. Overcoming existing obstacles and establishing internationally accepted certification standards is crucial for the continued growth and development of organic aquaculture.

Benefits of Organic Aquaculture in India

- Certain consumers opt for organic products because they contain fewer synthetic chemicals, are produced with consideration for animal welfare, and may be believed to offer enhanced nutritional value.
- Products derived from organic farming, free from synthetic chemicals, offer consumers a healthy and flavorful option, fostering a secure and sustainable environment.

Chapter-34

Recent Innovative Technologies in Aquaculture

Introduction

Over the next decade, aquaculture encounters various challenges, including disease control, enhancement of brood stock, development of suitable feeds and feeding methods, advancement of hatchery and grow-out technology, and effective management of water quality. These areas offer significant opportunities for interventions involving biotechnology and other technologies.

Some recent technological inventions are,

- Innovations in Reproduction
- Management of Diseases
- Technological Advances in Feeding
- Improvement of Holding Systems
- Advancements in Aquaculture Enhancement Technologies

Reproduction Innovations

The application of genetic principles to enhance production in aquatic animals currently lags behind advancements seen in the plant and livestock sectors. However, carp and tilapia culture in Asia are experiencing benefits from genetics research, particularly in genetic sequencing and the development of specific genetic markers. An example of this progress is evident in the GIFT (Genetic Improvement of Farmed Tilapia) project, which focuses on enhancing productivity in tilapia farming. Consumer concerns regarding the use of hormones in food animals are growing, and hybrid crosses that result in skewed sex distributions may not be optimal for farm productivity. Alternative methods for achieving monosex populations include cloning through nuclear transplantation and gynogenesis.

Disease Management

Shrimp broodstock management programs are currently focusing on two complementary objectives: the production of specific pathogen-free (SPF) stocks. These approaches are gaining traction in shrimp stocking initiatives across countries like the USA, Venezuela, and French Polynesia, utilizing species such as P. vannamei and P. stylorostris. While both strategies aim to produce "high health" stocks, it's noted that many SPF stocks exhibit subpar performance.

In the realm of disease control and prevention in shrimp aquaculture, DNA vaccines hold promising potential. Molecular probes have been commercially developed for detecting pathogens like IHHNV and type-A baculovirus, with ongoing efforts for the development of probes targeting other viral pathogens such as white spot, SEMBV, MBV, TSV, HPV, and YHV.

Feed Technologies

Feed Technologies include live feed culture and management. It includes ,

Live feed culture

- Artemia decapsulation technique
- Preparation and use of Artemia.

Feed management

- Microencapsulated feed
- Feed formulation method
- Bio-encapsulated feed.

Holding Systems

A notable technological advancement in the US freshwater farming sector is the substantial increase in tilapia production through indoor closed recirculation systems. Despite this progress, American production remains overshadowed by imports from countries like China, Costa Rica, Ecuador, and Honduras, where production can be achieved with lower capital investments.

Offshore cage development, observed in regions like the US, Hawaii, and India, involves the utilization of double cone-shaped nets suspended on a central floating vertical support column. These cages can be permanently submerged, with feed administered through a pipe from the surface.

Aquaculture Related Enhancement Technology

Fortunately, A variety of emerging and disruptive technologies hold the potential to transform the aquaculture industry. These technologies encompass...

Robotics

The application of robots extends to various tasks in aquaculture, including feeding, pond and net cleaning, vaccine injection, and the removal of sick fish. Additionally, robots are utilized for surveying fish health and preventing escapes from fish farms. Importantly, robots can enhance the profitability of aquaculture by operating continuously and independently, even in adverse environmental conditions, without requiring human intervention.

Drones

Drones are utilized for data collection, allowing for the monitoring of both land-based and offshore fish farms, including the inspection of cages for any holes or damages. Additionally, drones play a crucial role in gathering unique and hard-to-obtain information. For instance, Sail drone collects farm data, analyzes fish stocks, and tracks environmental conditions.

Sensors

Sensors play a crucial role in gathering various water quality parameters, such as dissolved oxygen (DO) levels, pH values, salinity, turbidity, and pollutant concentrations. In aquaculture, biosensors have been developed and utilized to analyze DO levels, water salinity, and temperature. In the salmon industry, individual heart rates and metabolism can be monitored and recorded. By employing underwater sensors connected to the internet, the feeding frequency and amount for cultured fish in cages, ponds, and rivers can be adjusted based on their hunger status. This optimized feeding approach can significantly improve feed utilization efficiency and reduce feed wastage, ultimately lowering total production costs. China has pioneered the development of deep-sea cages equipped with numerous sensors to monitor water quality, fish hunger status, net integrity, and fish movement. Looking ahead, there is a need to develop real-time sensors capable of measuring the stress levels of individual fish and detecting waterborne pathogens. These sensors should be easily deployable either within live fish or within the water, transmitting robust signals detectable by devices on land, boats, or satellites.

- ❑ Information/digital technologies
- ❑ Offshore farming
- ❑ RAS
- ❑ Replacement of fish meal and oils and oral vaccines.

The advent of sonar and GPS technologies has transformed fishing practices, allowing fishermen to pinpoint schools of fish with accuracy. This has resulted in notable reductions in search times and enhanced catch rates, ultimately yielding improved returns on investment

Glossary Terms

Aeration: The process of increasing oxygen levels in water using mechanical devices such as aerators, blowers, or paddlewheels, essential for supporting higher stocking densities and intensive production in semi-intensive aquaculture systems.

Aquaculture Economics: The branch of economics that focuses on the production, distribution, and consumption of aquaculture products, as well as the economic impacts and implications of aquaculture activities.

Aquaculture: The cultivation of aquatic organisms such as fish, crustaceans, mollusks, and aquatic plants under controlled conditions.

Biomass: The total weight of living organisms, usually expressed per unit area or volume, important for assessing production levels and carrying capacity in aquaculture.

Biosecurity: Measures implemented to prevent the introduction and spread of diseases and pathogens in aquaculture facilities, including quarantine protocols, hygiene practices, and control of external sources of contamination.

Bivalve: A class of mollusks characterized by hinged shells, including species such as mussels, oysters, and clams, commonly cultivated in aquaculture for their high protein content and economic value.

Broodstock: Mature organisms used for breeding purposes in aquaculture, selected for desirable traits such as growth rate, disease resistance, and size.

Carrying capacity: The maximum population size of a species that a given environment can sustain indefinitely without detrimental effects on the ecosystem.

Cost-Benefit Analysis: A method used to evaluate the economic feasibility of aquaculture projects by comparing the costs incurred with the benefits gained, typically expressed in monetary terms.

Dissolved oxygen: The amount of oxygen present in water, crucial for the survival of aquatic organisms, particularly fish.

Economies of Scale: The cost advantages gained when aquaculture production increases, leading to lower average costs per unit of output due to factors such as improved efficiency and utilization of resources.

Extensive Aquaculture: A type of aquaculture characterized by low stocking densities and minimal management inputs, relying primarily on natural productivity and ecosystem dynamics.

Externalities: Costs or benefits associated with aquaculture production that are not reflected in market prices, such as environmental impacts, social welfare effects, and ecosystem services.

Feed Conversion Ratio (FCR): A measure of the efficiency of feed utilization in aquaculture, calculated as the ratio of feed input to biomass gain, used to assess feeding strategies and optimize production efficiency.

Feeding Regimes: Scheduled feeding practices employed in semi-intensive aquaculture to supplement natural food sources and optimize growth rates, typically using formulated feeds containing balanced nutrients and vitamins.

Fertilization : The application of organic or inorganic fertilizers to stimulate natural productivity and plankton growth in extensive aquaculture ponds, enhancing water quality and providing supplementary feed for cultured organisms.

Fingerling: Juvenile fish typically larger than fry but still small and delicate, often raised in hatcheries before being transferred to grow-out facilities in aquaculture operations.

Fishery Economics: The study of economic aspects related to wild fisheries, including resource management, market dynamics, and policy interventions, often compared and contrasted with aquaculture economics.

Fishmeal: A protein-rich ingredient derived from processed fish or fish byproducts, commonly used in aquafeed formulations to provide essential nutrients and amino acids for fish growth and development.

Fry : Newly hatched or juvenile fish, typically at the early stages of development.

Harvesting: The process of collecting and removing aquatic organisms from extensive aquaculture ponds, typically conducted periodically based on market demand, growth rates, and production cycles.

Hatchery: A facility where fish eggs are artificially fertilized, incubated, and hatched under controlled conditions, often used in aquaculture for producing fry and fingerlings.

Input-Output Analysis: A method used to examine the interdependencies and relationships between different sectors of the aquaculture industry and the broader economy, focusing on inputs, outputs, and economic linkages.

Integrated Multi Trophic Aquaculture (IMTA): A sustainable aquaculture approach that involves cultivating multiple species at different trophic levels (*e.g.*, fish, shellfish, seaweeds) to optimize resource use and minimize environmental impacts.

Marginal Cost: The additional cost incurred by producing one more unit of aquaculture output, representing the change in total cost resulting from a small change in production level.

Mixed Farming: The integration of aquaculture with other agricultural activities, such as rice farming or livestock production, to optimize resource utilization, nutrient cycling, and overall farm productivity.

Monosex culture: Aquaculture production consisting entirely of a single sex, usually achieved through sex reversal or hormonal manipulation to promote faster growth and reduce competition.

Net Present Value (NPV): A method used to assess the profitability of aquaculture investments by discounting future cash flows to their present value, considering the time value of money.

Overfishing: The harvesting of fish stocks at a rate exceeding their natural reproduction and growth rates, leading to depletion of populations and ecosystem imbalance.

Polyculture: The simultaneous cultivation of multiple species in the same aquaculture system, aimed at maximizing resource utilization, enhancing ecosystem stability, and diversifying production outputs.

Pond Culture: A common form of extensive aquaculture where aquatic organisms, such as fish or shrimp, are raised in earthen or man-made ponds, utilizing natural or supplemented feeds and minimal intervention.

Price Elasticity of Demand: A measure of the responsiveness of aquaculture product demand to changes in price, indicating the percentage change in quantity demanded for a given percentage change in price.

Profit Maximization: The objective of aquaculture firms to maximize their economic profits by producing at a level where marginal revenue equals marginal cost, considering factors such as market demand and production costs.

Raceway Systems: Flow-through aquaculture systems consisting of elongated channels or raceways through which water is continuously circulated, providing oxygenation, waste removal, and environmental control for cultured organisms.

Recirculating Aquaculture System (RAS): A closed-loop aquaculture system that recycles and treats water within the facility, minimizing water usage and environmental impacts.

Salinity: The concentration of dissolved salts in water, measured in parts per thousand (PPT) or practical salinity units (PSU), an important parameter in aquaculture for species selection, water quality management, and site suitability assessments.

Selective Breeding: The controlled mating and genetic selection of brood stock in semi-intensive aquaculture to improve desirable traits such as growth rate, disease resistance, and fillet quality in cultured populations.

Semi-Intensive Aquaculture: A hybrid approach combining elements of extensive and intensive aquaculture, characterized by moderate stocking densities, supplementary feeding, and some level of management and infrastructure.

Stocking Density: The number of aquatic organisms per unit area or volume of culture system in semi-intensive aquaculture, carefully managed to balance growth, water quality, and environmental impacts.

Stocking density: The number of organisms per unit area or volume in an aquaculture facility, often optimized to maximize production while maintaining water quality and animal welfare.

Tank Culture: Aquaculture practiced in tanks or containers, allowing for greater control over water quality, feeding, and environmental conditions compared to extensive pond systems, commonly used for raising fish, shrimp, or other aquatic organisms.

Tilapia: A group of freshwater fish species belonging to the cichlid family, widely cultivated in aquaculture due to their rapid growth, tolerance to varying environmental conditions, and versatile culinary qualities.

Trophic level: The position of an organism in a food chain, determined by its feeding habits and nutritional source. In aquaculture, understanding trophic levels helps in designing efficient feeding strategies and managing nutrient cycling.

Water Exchange: The process of replacing water within aquaculture systems to maintain optimal water quality parameters, including removal of waste products, replenishment of oxygen, and control of temperature and salinity.

Water Management: Techniques and practices used to regulate water levels, quality, and flow within extensive aquaculture systems, including pond construction, water exchange, and drainage.

Water Quality Monitoring: Regular assessment of key parameters such as dissolved oxygen, pH, ammonia, and nitrite levels in semi-intensive aquaculture systems, crucial for maintaining optimal conditions for growth, health, and productivity.

Water quality: The physical, chemical, and biological characteristics of water, including parameters such as temperature, pH, dissolved oxygen, ammonia, and nitrate levels, crucial for the health and growth of aquatic organisms.

Yield: The quantity of aquatic organisms harvested from extensive aquaculture ponds, often measured in weight or volume per unit area of pond surface, influenced by stocking density, feeding practices, and environmental conditions.

Zooplankton: Small aquatic organisms, including tiny animals and larvae, that drift with water currents and serve as food for many fish and other aquatic organisms in aquaculture systems.

References

Bag, N., Moulick, S. and Mal, B.C., 2017. Evaluation of economic viability and sensitivity analysis of Indian major carp culture. *Indian Journal of Fisheries, 64*(3).

Beveridge, M.C.M. (Ed.),2004. Cage Aquaculture, 3rd Edition. Blackwell Publishing. Boyd, C.E.,1995. Bottom soils, sediments and pond aquaculture. Springer Publishing

Burnell,G. and Allen, G.,2009. New technologies in Aquaculture.1st Edition. Woodhead Publishing House.

Chakrabarti,N.M.,2012. Biology, culture and production of Indian major carps. Narendra Publishing House.

Christenson K.,2015.Aquaponics: Aquaculture- An introduction to aquaculture for small farmers

De Silva,S.S and Anderson, T.A.,1995. Fish Nutrition in Aquaculture. Chapman & Hall Aquaculture Series.

De, H.K., Shasani, S., Das, M.K. and Saha, G.S., 2020. An Economic Appraisal of Composite Carp Culture in West Bengal. *Journal of Extension Education, 32*(2), pp.6492-6499.

Edwards, P., 1998 .A systems approach for the promotion of integrated aquaculture. *Aquaculture Economics & Management, 2*(1), pp.1-12.

Jhingran, V,G.,1991.Fish and Fisheries of India.3rd Ed. Hindustan Publishers.

Little,D. and Edwards, P.,2003. Integrated livestock-fish farming systems. FAO Publications. Lovell,R.T., 1998. Nutrition and Feeding of Fishes. Chapman & Hall.

Nisar, U., Peng, D., Mu, Y. and Sun, Y., 2022. A solution for sustainable utilization of aquaculture waste: a comprehensive review of biofloc technology and aquamimicry. *Frontiers in Nutrition, 8,* p.791738.

Pillay, T.V.R.,1990. Aquaculture: Principles and Practices. Fishing New Books, Cambrige University Press Cambridge.

Pinho, S.M., de Lima, J.P., David, L.H., Emerenciano, M.G., Goddek, S., Verdegem, M.C., Keesman, K.J. and Portella, M.C., 2022. FLOCponics: The integration of biofloc technology with plant production. *Reviews in Aquaculture, 14*(2), pp.647-675.

Raizada, Sudhir & Pravin, P. & Jena, J.K.. (2019). ICAR Technologies: Breeding and seed production of finfishes and shellfishes.

Rath, R.K.,2000. Freshwater Aquaculture. Scientific Publishers.

Ross, L.G., Telfer, T.C., Falconer, L., Soto, D., Aguilar-Manjarrez, J., Asmah, R., Bermúdez, J., Beveridge, M.C.M., Byron, C.J., Clément, A. and Corner, R., 2013. Carrying capacities

and site selection within the ecosystem approach to aquaculture. *Site selection and carrying capacities for inland and coastal aquaculture, 19.*

Ruby, P., Ahilan, B., Antony, C. and Selvaraj, S., 2022. RECENT TRENDS IN AQUACULTURE TECHNOLOGIES. *Journal of Aquaculture in the Tropics, 37*(1-4), pp.27- 34.

Sasikumar, G. and Viji, C.S., 2015. Integrated Multi-Trophic Aquaculture Systems (IMTA).

Shafeena, T., 2016. Smart aquaponics system: Challenges and opportunities. *European Journal of Advances in Engineering and Technology, 3*(2), pp.52-55.

Soderberg, R.W.,1995. Flowing water fish culture, Lewis Pblishers.

Thomas, P.C.,Rath, S.C.and Mohapatra, K.D.,2003. Breeding and Seed Production of Finfish and Shellfish, Daya Publishing House, New Delhi.

Tidwell, J.H.(Ed.),2012. Aquaculture Production Systems. Wiley- Blackwell Publishing Yoram, A.,2015. Biofloc technology: a practical guidebook. WAS Publications.